# Pollution and Atmosphere in Post-Soviet Russia

# Pollution and Atmosphere in Post-Soviet Russia

## *The Arctic and the Environment*

Lars Rowe

**BLOOMSBURY ACADEMIC**
LONDON · NEW YORK · OXFORD · NEW DELHI · SYDNEY

BLOOMSBURY ACADEMIC
Bloomsbury Publishing Plc
50 Bedford Square, London, WC1B 3DP, UK
1385 Broadway, New York, NY 10018, USA
29 Earlsfort Terrace, Dublin 2, Ireland

BLOOMSBURY, BLOOMSBURY ACADEMIC and the Diana logo
are trademarks of Bloomsbury Publishing Plc

First published in Great Britain 2021
This paperback edition published in 2022

Cover design by Adriana Brioso
Cover image: Vasili Ragulin, 43 at the time, poses in the Pechenganikel smelter,
December 1995. (© Ola Solvang)

A catalogue record for this book is available from the British Library.

A catalogue record for this book is available from the Library of Congress.

ISBN: HB: 978-0-7556-0047-2
PB: 978-0-7556-3489-7
ePDF: 978-0-7556-0048-9
eBook: 978-0-7556-0049-6

Typeset by Deanta Global Publishing Services, Chennai, India

To find out more about our authors and books visit www.bloomsbury.com and
sign up for our newsletters.

*For Vera and Isak*

# Contents

# Illustrations

## Map

## Figure

## Photos

# Acknowledgements

Thanks to the Fridtjof Nansen Institute, and more specifically now former director Geir Hønneland for supporting this work, particularly through less productive periods. Thanks to Ola Solvang, who has allowed me to use some of his remarkable images from Nikel. Thanks also to Amund Trellevik and Anne Berteig for letting me use their photos. Professor Sven Holtsmark has been an invaluable resource throughout my career. This study was no exception. I am also indebted to Magne Røed and Jan Thompson, both now retired from their positions in the Norwegian Ministry of the Environment. Magne, Jan and their colleagues at the ministry archive were most helpful in making documentation available to me.

Thanks lastly, to my family. I dedicate this book to Vera and Isak, in the hope that they will experience a future less endangered by waste from human activities.

Hølen, spring 2020
Lars Rowe

The research for this book was funded by the
Norwegian Ministry of Defense.

# Abbreviations, acronyms and Russian terms used

AIP (in notes)  Miljøverndepartementets avdeling for internasjonalt miljøvernsamarbeid og polarsaker (Department for International Environmental Cooperation and Polar Issues in the Norwegian Ministry of the Environment)

BEAR  Barents Euro-Arctic Region

ET (in notes)  Elkem Technology

GIEK  Garanti-instituttet for eksportkreditt (the Norwegian Export Credit Agency)

Goskomekologiya  Gosudarstvennyi komitet po ekologii (State Committee for Environmental Protection)

Goskomgidromet  Gosudarstvennyi komitet po gidrometeorologii i monitoringu okruzhayushchei sredy (State Committee for Hydrometeorology and Environmental Monitoring)

Goskompriroda  Gosudarstvennyi komitet po okhrane prirody (State Committee for Environmental Protection)

KGMK  Kolskaya Gorno-Metallurgicheskaya Kompaniya (Kola Mining and Metallurgical Company)

KMO  Komitet Molodezhnykh Organizatsii (Committee for Youth Organizations)

M. (in notes)  mappe (Norwegian archival unit (file))

| | |
|---|---|
| MD (in notes) | Miljøverndepartementets arkiv (Archive of the Norwegian Ministry of the Environment) |
| Minpriroda | Ministerstvo po okhrane prirody (Ministry for Environmental Protection) |
| NIB | Nordic Investment Bank |
| NME | Norwegian Ministry of the Environment |
| NOK | norske kroner (standard abbreviation for the Norwegian currency, kroner) |
| oblast | Soviet/Russian administrative unit, county |
| PERG | Pechenga Nickel Expert Review Group (est. 2000 to assess the Russian modernization proposal involving Vanyukov furnaces) |
| PRC | Pechenga Reconstruction Consortium (unit created by Norwegian Elkem Technology, Norwegian Kvaerner Engineering and Swedish Boliden Contech, for modernization of Pechenganikel installations) |
| raion | Soviet/Russian administrative unit, municipality |
| RSFSR | Rossiiskaya Sovetskaya Federativnaya Sotsialisticheskaya Respublika (Russian Soviet Federative Socialist Republic) |
| SFT | Statens forurensingtilsyn (Norwegian Agency for Pollution Control) |
| SMK (in notes) | Statsministerens kontor (Office of the Norwegian Prime Minister) |
| Tsvetmetekologya | Ecological branch of the Soviet Ministry for Non-ferrous Metallurgy |
| zapovednik | nature protection area |

# 1

# Introduction

*But nature has little time to lose. We are still closer to where we started than we are to the goal line. The reconstruction of the nickel works is a gigantic industrial project, the Soviet Union is going through a deep transformation. Responsibilities are unclear. Decision making processes in the east are difficult, but decisions are necessary. Time is of the essence.*

Jan Peder Syse, 1990[1]

While the small Norwegian municipality Sør-Varanger, in the county of Finnmark, is unique in many ways, its primary claim to fame lies in its location. The eastern stretches of Sør-Varanger's territory make up Norway's national border with Russia. This fact has a multitude of implications for Sør-Varanger's municipal administrators, local NGOs and roughly 10,000 inhabitants. The main settlement Kirkenes has a palpable Russian presence, and sports clubs, various cultural groups and other associations are engaged in a plethora of collaborative projects with Russian counterparts.

However, the proximity to Russia does not only produce cooperation, trade and cultural exchange. It can at times, especially during the cold Arctic winter when meteorological factors conspire to put a frosty lid over the area, result in something entirely different. For instance, on 25 February 2019, during an especially frigid period when temperatures plummeted towards minus 40 degrees Celsius, the Sør-Varanger administration felt obliged to warn local citizens to stay indoors. It was not the freezing cold that constituted the danger (people in Finnmark are well equipped to handle low temperatures). Rather, it was a specific Russian export which flowed unchecked across the

The Pechenganikel plant seen from the southeast and facing the Russian–Norwegian border. Note the extensive damages to vegetation.
Photo: Amund Trellevik

Russian–Norwegian border that necessitated this public announcement: streams of sulphur dioxide emitted from the industry in the Russian border town Nikel. According to the local newspaper, an air quality measuring station close to the national border reported sulphur dioxide content in the atmosphere that well surpassed the established danger levels of 500 milligram per cubic metre of air.

The same article informed that the Russian owner-oligarch Mikhail Potanin pledged to reduce his company's total emissions by 75 per cent by 2023.[2] Potanin's statement, promising air quality improvement around his industrial facilities, is but one of many that have been issued by him and previous owners over the years. To the inhabitants of Sør-Varanger, such promises will hardly raise any hopes. They have seen bright prospects fall through on too many occasions to be swept away by optimism.

In this book, we shall look closer at the historical processes that so persistently curb their enthusiasm by examining the many attempts at decreasing industrial pollution stemming from the Pechenganikel plant in Russia's northwestern corner. These efforts have since the late 1980s involved, inter alia, Soviet and Russian industry and government agencies, Finnish and Swedish industrialists and officials and, not least, Norwegian environmental authorities, foreign services and ecological activists. Thus, the history of pollution control in Northwest Russia is both a transnational history and an environmental history. It is also, inasmuch as the book follows the development from the dying breaths of the Soviet Union to the final Norwegian decision to withdraw from the collaborative efforts in 2010, a contribution to post-Soviet history. Finally, this book is an account of how Russia's transformation from a socialist superpower to a quasi-democratic and ultra-capitalist state was handled by Western societies, in this case a small neighbouring country.

Before we embark on the main narrative of this book, I will introduce some background information and analytical approaches that will help frame the empirical investigation. One of the more astounding features of the thirty-year history of the Pechenganikel modernization is that the collaborative efforts to curtail polluting emissions have barely made a dent in Pechenganikel's practices. This suggests that there are deeper constraints than just scarce funding and technological difficulties at play. In the coming section, I argue that one main impediment to success lay in how industrial activity and industrial waste was and is perceived of in the Soviet Union and later Russia. I then round off this introduction by briefly outlining two approaches to understanding ideas, interests, motivations and restraints that surround international relations. These approaches will be revisited in the concluding chapter of this book, and thus inform my analysis of why and how the Pechenganikel modernization never materialized.

## Soviet and Russian environmental thinking

In the final years of the Soviet Union and in the aftermath of its collapse, environmental issues were in vogue, for various reasons. The attention directed at Soviet environmental practices, or lack thereof, was immense.

Exposed to the outside world, Soviet and later Russian authorities came under pressure to deal with the problems of industrial discharges that had apparently been ignored during seventy years of Soviet rule. In some instances, Soviet pollution was harmful not only to Soviet/ Russian territory, but also to neighbouring countries. Pechenganikel was an example of this, and its emissions became a contentious question in Soviet and later Russian relations with the Nordic neighbours. As background to my treatment of the meeting between Soviet and Nordic (predominately Norwegian) ideas and interests that took place in the Pechenganikel question, some aspects of the scholarly debate on Soviet approaches to nature and pollution merit attention here.

Recent research has aimed to nuance the commonly promoted image of Soviet society as one-sidedly preoccupied with the wealth-accruing potential of nature and natural resources. Attention to nature, it is argued, was an accompanying feature of Soviet industrialization, though admittedly one that rarely enjoyed priority status. As will be maintained in the following, the insights provided by these studies, however valuable, do not provide the basis for a revision of the basic understanding that the Soviet state (and, by extension, its Russian successor) invariably ranked industrial interests over environmental values.

In his comprehensive historical study of the Soviet conservationist movement, Douglas R. Weiner shows how the existence of this essentially oppositional grouping was accepted by the Soviet leadership. Finding no apparent answer as to why this came to be, he arguably, and perhaps inadvertently, provides the answer, in pointing to the movement's 'single-minded focus on the protection of "pristine" nature'. The movement never truly challenged industrial interests, but rather served to uphold a division between industrialized areas (where acceptance of pollution was high) and wilderness areas that were protected as *zapovedniki* (where pollution or other human influences were kept to a minimum). In Weiner's interpretation, it seems, the main significance of the Soviet environmental movement did not reside in its ability to fight pollution, but in its role as an accepted vehicle for (minor) dissent. For Soviet conservationists, 'environmental activism provided the feeling (and sometimes the fact) that they were tangibly and independently defending the good of the community in the face of a repressive, wasteful, and destructive bureaucratic system.[3]

A social science study of the Soviet environmental legacy addresses what is seen as a knowledge gap among non-Russian researchers in the aftermath of the Soviet collapse. Pointing to what he calls 'a number of underlying preconceptions and biases', Jonathan D. Oldfield criticizes Western scholars' ignorance of Soviet nature protection agencies and history of environmental management. Drawing inter alia on Weiner's work, Oldfield traces the roots of the conservationist movement back to the pre-revolutionary era and describes the network of protected *zapovedniki*. Furthermore, he writes, the Soviet regime 'left a reasonably extensive environmental monitoring infrastructure in addition to systems of natural resource management and environmental impact assessment'. He concludes that 'a more sensitive understanding of Russia's environmental legacies is essential if we are to move away from the notion of Soviet society possessing a limited social and intellectual capital in the general area of environmental protection'.[4]

Like Weiner and Oldfield, historian Andy Bruno identifies a Soviet reverence for nature. In his book *The Nature of Soviet Power*, Bruno argues that varying degrees of holistic approaches accompanied the Soviet inclination to view nature as an impediment that should be conquered for the sake of human betterment. While clearly poorly protected, the natural environment was not disregarded. Bruno argues that Soviet industrialism, in its essence, was not uniquely antagonistic to the natural world. Rather, Soviet industry related to resources much the same way as its capitalist counterparts and viewed the riches of the natural world as wealth-accruing substances to be exploited. There existed, however, an undeniable gap between environmental standards in the Soviet Union and large parts of the capitalist world towards the end of the Soviet era. Bruno, acknowledging this, seems to ascribe the prolific environmental degradation that became evident in the Soviet Union from the mid-1980s onwards mostly to a lack of flexibility in the planned economy in the face of global economic shifts in the 1970s and 1980s.[5] I will, on the basis of the empirical study in this book, argue that such an exogenous explanation cannot fully account for the remarkably damaging environmental practices of Soviet industry. Rather, I emphasize a factor indigenous to the Soviet, and later Russian, system: that national industry, unlike its Western counterpart, was able to operate unchecked by any form of restrictive and empowered environmental agencies.

Neither Weiner nor Oldfield nor Bruno argue that Soviet industrialization was anything less than disastrous for the natural landscapes that were exposed to it, but their main objective has been to identify traces of Soviet environmental concerns. Consequently, their studies tend to emphasize evidence of such tenets. The purposes of the present study are quite different. Unlike these authors, I see Soviet and Russian environmental policies in the contrasting light cast by neighbours who held different views about industrial pollution and its detrimental effects on the environment. I will argue that the *actual impact* of environmental concerns on Soviet and later Russian policies can be assessed better in this comparative light. In other words, while Weiner, Oldfield and Bruno have demonstrated that the environment did matter to the Soviet state, the question remains exactly *how* and *to what degree* it mattered.

The historical trajectory that culminated in the establishment of modern and influential environmental protection agencies as seen in many Western countries from the early 1970s was not paralleled in the Soviet Union.[6] Oldfield does demonstrate that evidence of environmental concerns in government can be traced continuously throughout Soviet history, but he is acutely aware that these concerns never trumped the logics of productivity in the centrally planned economy. Rather than being a policy area in their own right, environmental issues relating to industrial production were dealt with (or ignored) by the various Soviet branch ministries. Only in 1988 did the Soviet Union get its first central (union-level) agency for environmental protection – the State Committee for Nature Protection (*Goskompriroda*).[7]

In the Soviet system, there was always a strong inclination to see problems of pollution as a matter for industry itself, and not as an issue for general political debate. During the 1970s, a range of regulations aimed at protecting water, minerals, forests and air quality were enacted in Soviet legislature. Oldfield points to the 1972 Supreme Soviet decree 'Concerning Measures for the Future Improvement of Nature and the *Rational Utilization of Natural Resources* [italics added]' as the forerunner for this legislative process.[8] The title of the decree is in itself a pointer towards Soviet, and later Russian, approaches to industrial pollution. It couples the ambition to protect the environment with 'rational utilization of natural resources', reflecting the belief that these two goals were seen as symbiotic rather than opposed to each other. The best way to protect the natural environment from industrial waste was not to set limitations for polluting activities, but to strive to optimize industrial processing of raw material.

The emphasis on advances in industrial technology and methods for exploitation of raw materials to reduce pollution is not exclusive to Soviet approaches to environmental problems. Enhancing the environmental standards of production processes is a widespread and obvious ambition in Western countries as well. The Soviet uniqueness resides in the understanding of pollution as a largely acceptable side-effect of the all-important industrial production. Rather than it representing an intolerable damage to the natural environment, and in some cases to human health, pollution was expected to disappear once the industry found better ways of refining its raw materials. Pollution was a purely industrial concern, not a mainstream political matter; and all aspects of industrial activity, including harmful discharges, were to be handled by the industry itself, through unceasing efforts to enhance production methods. The habitual reference to the need to optimize the exploitation of natural resources is evident not only in the abovementioned Supreme Soviet decree from 1972, but in many subsequent Soviet and Russian statements and documents pertaining to industrial pollution.[9] Full, or complex, utilization of raw materials was as a crucial tenet in Soviet and later Russian thinking about the problem of industrial waste.

The concept of 'complex utilization' was developed in the 1930s by the prominent Soviet geochemist Aleksandr Fersman (1883–1945). Fersman's programme was intrinsically bound to a deeply utilitarian understanding of nature.[10] He argued that the 'complete use of all mining mass extracted from the earth' could be made possible through technological advances and comprehensive recycling of the slag resulting from the refining process. He not only was very hopeful of heightened productivity and efficiency, but also enthusiastically foresaw that such a 'complex utilization' of mineral ore would mark the end of pollution, as 'nothing is emitted into the air and washed away by water' – it would all be used.[11]

Though arguably utopian, Fersman's ideas came to influence Soviet governance and industry, forming an important part of the country's environmental tradition. Pollution problems were simply ascribed to the suboptimal utilization of resources and understood as an expression of the still-underdeveloped production system in the Soviet Union. The concept of 'complex utilization' became institutionalized in the Soviet system and was later referred to by civil servants and politicians in post-Soviet Russia. As an idea that shaped Soviet and Russian policy choices, it was very different from the typical Western inclination to see industrial activity and environmental protection as two conflicting aims. I will in this study argue that the contradiction between Soviet/ Russian and Norwegian views was a core problem in the Pechenganikel negotiations.

To sum up, in Soviet and Russian approaches to the environment there emerged a duality that would cast long shadows over post-Soviet Russia, with a sharp divide between classical conservationism and modern environmentalism. On the one hand, there was the conservationist movement which focused on landscape protection. This did not interfere with Soviet industrial interests and was therefore allowed to exist. The ultimate goal of the conservationist movement, the protected landscape area or *zapovednik*, was 'an inviolable nature reserve dedicated to long-term scientific, especially ecological, study'.[12] These physically closed-off *zapovedniki* were not open to the general public, but functioned as museum-like pockets where pristine nature could be observed by qualified individuals only. Consequently, conservationism became an elitist preoccupation which never engaged the general public.

On the other hand, modern environmentalism, understood as activism or governance practices directed against polluting industry, barely existed. There was little or no pressure on Soviet industry to curtail its emissions. Industrial pollution became a problem for the industry to solve by itself. Branch ministries would, to the extent this did not conflict with overriding productivity goals, seek to alleviate the contaminating effects of production on nature through technical improvements in line with the ideal of 'complex utilization' of natural resources. In the Soviet planned economy, however, industry focused on production targets, employment and social services. In this hierarchy of problems, industrial pollution would always be at the bottom of the list.

Unlike in Western countries, there was never an amalgamation between classical conservationism and modern environmentalism in the Soviet Union. Thus, a holistic understanding of nature as an entity worthy of protection both within and outside the borders of designated wilderness areas never truly materialized. What little there was of Soviet and later Russian environmentalism did not mature into a sustained popular movement that eventually forced its way into governmental offices. Except for a brief but significant period around the Soviet collapse in 1991, when environmental movements served as vehicles for public discontent, ecological concerns never came to loom large in Soviet or Russian public opinion or government corridors.

Thus, in raising questions about the pollution problem in Pechenga, Nordic officials found themselves facing a political reality fundamentally different from their own. The chasm between the two countries' ideas on handling industrial pollution was deep, and bridging this gap would always be essential for the eventual success of efforts to curtail discharges from Pechenganikel. In this respect, the Pechenganikel modernization scheme discussed in the present study became a touchstone for post-Soviet Russian–Norwegian relations: could such conflicting ideas be reconciled?

## Ideas and interests in foreign policy

The Pechenganikel modernization negotiations to a large degree became a foreign policy arena for contrasting Russian and Norwegian ideas about handling industrial pollution. I will in the following present a conceptual framework that will help in analysing this clash of ideas. However, acknowledging the rather obvious fact that not only ideas but also *interests* tend to be pivotal to the shaping of foreign policy, I will thereafter briefly discuss an analytical model that provides a vocabulary for discussing interests.

Writing on the role that ideas might play in international politics, Judith Goldstein and Robert O. Keohane have become involved in a theory-skirmish with their fellow social scientists about the relative impact that interests and ideas have on policy formation. Goldstein and Keohane argue forcefully that political choices are *not* made solely on the basis of the immediate interests of a given political individual or institution, but that a full explanation of policy formation must also reflect the underpinning beliefs of said political individual or institution.[13] They criticize 'approaches that deny the significance of ideas' while not challenging 'the premise that people behave in self-interested and broadly rational ways'.[14] To most historians, who are regularly confronted with empirical evidence that speaks to the complexities of human interaction, there seems to be little here to get worked up about – both ideas and interests obviously matter in policy formation as well as in foreign politics (and, of course, ideas are part of the formation of interests).

That said, in setting up their analytical framework, Goldstein and Keohane move on to a more illuminating and promising discussion.

They describe a hierarchical model of ideas, at the top of which are *world views*. World views, they explain, are 'embedded in the symbolism of a culture and deeply affect modes of thought and discourse'.[15] Among the examples mentioned are the world religions, adherence to human rights principles or state sovereignty, and Stalinism. Below world views in this hierarchy we find *principled beliefs*. These are often, but not always, justified in terms of larger world views: they are normative ideas of what is right and what is wrong. Lastly, there are *causal beliefs*: 'beliefs about cause–effect relationships which derive authority from the shared consensus of recognized elites, whether they be village elders or scientists at elite institutions'.

Goldstein and Keohane also bring in the temporal dimension by applying the term *institutionalization*. When ideas have become accepted and referred to over a long period of time, they lead to lasting changes in existing rules and norms – they become institutionalized:

> Although the institutionalization may reflect the power of some idea, its existence may also reflect the interests of the powerful. But even in this case, the interests that promoted some statute may fade over time while the ideas encased in that statute nevertheless continue to influence politics. Thus at a later time, these institutionalized ideas continue to exert an effect: it is no longer possible to understand policy outcomes on the basis of configurations of interests and power alone.[16]

Thus, institutionalization allows ideas to survive within a political system even though the circumstances and the power structures that once brought them about may no longer be present. With that, policy formation is removed even further from the purely interest-driven realm. As I will argue in the concluding chapter, all three levels in this hierarchical model of ideas were factors in Russian–Norwegian Pechenganikel negotiations.

Despite the usefulness of this model, an analysis of a foreign policy process that fails to include an examination of the interests at play would be incomplete. A second conceptual model for understanding the multilayered nature of how domestic groups can affect international diplomacy offers a useful vocabulary. A few select elements borrowed from Robert D. Putnam's game-theoretical approach for analysing interest-driven international negotiations, what he calls *two-level games*, should help shed light on the largely empirical observations made in this book.[17]

Again, a rather obvious observation – that the domestic dimension of any international negotiation should not be ignored – is taken as the point of departure. Briefly put, Putnam's model of two-level games mirrors the intricacies and possibilities arising in many international negotiations because of the need, on both sides of the table, to take domestic interest groups into account. The increased complexity that arises from this can be a stumbling block for negotiators, but politicians and diplomats can also make use of the pressures in both their own and in the counterpart's domestic playing field when trying to maximize the results of international negotiations. Before the Pechenganikel negotiations started in earnest, several actors on both sides entered the frame. Representing varying and sometimes opposing interests, these groups attempted to sway their respective governments in their own direction, thereby preparing the ground for two-level games. Putnam's model refutes the idea that states are to be understood as unitary actors and asserts that the negotiators' room for manoeuvring in any bilateral or multilateral setting is limited by a patchwork of cross-cutting domestic interests.

The range of possible negotiation outcomes, Putnam argues, is limited by what types of international agreements are deemed tolerable in the domestic playing field. In other words, international negotiators are restricted to what they think is acceptable to interest groups, the bureaucracy, political actors and others at home. Putnam calls the range of acceptable negotiation outcomes the 'win-set'. The larger the win-set is perceived to be, the greater the chances of reaching agreement. Thus, negotiators who think that domestic interests are ready to accept a wide range of possible agreements, that is, that the win-set is large, will be correspondingly unrestricted in the negotiations.

However, at the negotiation table, negotiators will be aware of their opponents' presumed win-sets and will use this to their advantage. As Putnam puts it, 'The larger the perceived win-set of a negotiator, the more he can be "pushed around" by the other . . . negotiators.' Negotiators can, by referring to the large win-set of their counterpart, insist that their proposals are indeed acceptable to most interest groups in the opponent's home country. Conversely, Putnam argues, a negotiator can use a small win-set (or a largely negative array of domestic interests) to his or her advantage – by simply refusing to agree to make concessions, while pointing to the impossibility of obtaining acceptance at home.[18]

Although Putnam further elaborates his theoretical framework, we will leave him at this point. For the purposes of this study, the basic concepts of two-level games and win-sets are adequate for providing a relevant conceptual framework for analysing the Soviet/Russian–Norwegian negotiations about Pechenganikel's modernization. Suffice it to say here that any expectations of the Soviet Union or post-Soviet Russia as unitary rational actors are utterly negated by the empirical findings presented in this study. We shall return to this discussion in the concluding chapter of this book.

\* \* \*

Chapters 3 to 5 in this book are devoted to the period between 1988 and 2010. The development in Norwegian–Russian negotiations aimed at curtailing emissions from the Pechenganikel plant is thus traced over a period that covers three distinctly different periods in recent Russian history. In Chapter 3, the immediate backdrop for the negotiations was the rapid decline in Soviet statehood, which accelerated towards the end of the 1980s and culminated in the biggest upheaval in international politics since the Second World War: the Soviet collapse. Parallel to this development, the Norwegian government entered an environmental agreement with the Soviet Union in 1988 and re-entered a mostly unchanged agreement with the Russian Federation in 1992.

The intervening years offered little or no stability and very little room for effective implementation of environmental measures.

In Chapter 4, the Pechenganikel modernization schemes were surrounded by a political reality that was equally unstable, but for different reasons. Russia's first post-Soviet decade was haunted by an incessant internal struggle for economic and political power. It is in most countries and periods difficult to discern exactly where the economy stops and where politics begin. However, the chaotic Russia of the 1990s can serve as a particularly stark example of how political and economic elites can become interconnected and to a large degree overlap. Perhaps the most telling example of this was Boris Yeltsin's re-election in 1996, which would have been unthinkable had it not been for a massive campaign designed and implemented by the leading Russian tycoons – the oligarchs – and the media channels they controlled. In this period, which I argue amounted to a *state of emergency* in Russian politics, both internally and externally, the Pechenganikel modernization had little chance of success.

But Russian *normality* was restored. When Vladimir Putin was appointed and later elected president as the new millennium commenced, Russian statehood, politics and economic life regained its equilibrium. Boosted by extreme state revenues, Putin was able to re-establish the Kremlin as Russia's unopposed centre of power. While able to continue their economic activities, the oligarchs became politically sidelined. In Chapter 5, we look at how this new development was mirrored in the Pechenganikel modernization. On the one hand, the re-established stability made negotiations more viable. On the other hand, there was little to suggest that the reshuffling of political and economic power had changed underlying ideas of Russian statehood, not to mention Russian ideas about the relationship between industrial activities and their environmental effects.

This book starts out, however, with an introductory Chapter 2, which gives an overview of the industrial development in Pechenga from the early 1950s to the mid-1980s. During these thirty-odd years, industrial practices at the combine prepared the ground for the Pechenganikel modernization, as Russia's Nordic neighbours, especially Norway and Finland, were growing increasingly worried about the industrial pollution that was carried by air currents onto their territories. Parallel to Pechenganikel's industrial expansion in this period, environmental concerns and cross-border pollution became increasingly important in international politics. Thus, the next chapter also deals with how the Kremlin made use of such tenets in their efforts to further Soviet objectives on the international stage, which were mostly unrelated to the state of the environment.

# The advent of an environmental disaster

In the Norwegian national budget for 2009, a long passage concerning the environmental situation in and around Pechenganikel contains the following sentences: 'At the present, modernization of the smelting plant in Nikel seems to be out of the question. There is not enough time to complete a modernized installation that meets the requirements of agreements with Norway by 2010.'[1] Two years later, in the national budget for 2011, the following statement occurs:

> It is now clear that the modernization project in the nickel industry at the Kola Peninsula agreed between the Norwegian and Russian governments in 2001 will not go ahead as planned. . . . The agreement providing for Norwegian contributions has not been honored. Further contributions will not be paid out. The Norwegian authorities will make efforts to have earlier financial support reimbursed [from the Russians].[2]

How did Pechenganikel find its way into the national budget of a neighbouring state? As the quotations above indicate, certain agreements between Norway and Pechenganikel about modernization of the nickel works existed, although the conditions in these agreements were apparently not met. The details of how all this came about form the central theme in this book.

Nikel and its surroundings, like many other commercial regions in the Soviet Union, were by the end of the Cold War in an increasingly precarious environmental state after decades of industrial expansion. In the case of Pechenganikel, easily transported emissions from the combine's smokestacks were not contained within the country of origin. Due to its location close to the borders with Norway and Finland, pollution from Pechenga became a concern beyond Soviet borders.

Before we delve into Pechenganikel's last days as a Soviet enterprise and its post-Soviet existence, some key moments in the period between the first post-war years and the Soviet collapse merit attention. Instead of attempting a strict chronology, I will trace some development lines that point towards the reality in which the Pechenga area found itself towards the end of its Soviet existence. From being a highly industrialized, securitized and closed border zone, Pechenga became transformed into one of several areas that openly exhibited Soviet shortcomings in environmental protection. This chapter covers main developments in Pechenga from the mid-1950s to the mid-1980s, including the expansion and growth of the industrial complex and its adjacent urban settlements. We will also look briefly at the environmental consequences of industrial expansion in the Soviet Union, and the political approaches to international environmental cooperation in the Brezhnev and Gorbachev eras. All this prepared the ground for a strong international interest in Pechenga's nickel industry in the waning days of the Soviet Union and the early period of Russia's post-Soviet existence.

## Urbanization, industrialization and electrification

After the first turbulent and frantic years following the Soviet takeover of Petsamo in 1944, Pechenganikel entered a long period of steady growth. The massive nickel ore was continuously explored, and several new exploitable deposits were discovered by Soviet geologists. Two of these, the Pilgujärvi deposit (Soviet name Zhdanovskii, later Tsentralnyi) and the Allarencheskii deposit, gave impetus to the establishment of two new villages in Pechenga: Zapolyarnyi (1955–57) and Prirechnyi (1962) (see map 1).

**Map 1** The Pasvik valley and surroundings

Zapolyarnyi was truly a 'Soviet' village, in that it was mostly built by 'young enthusiasts' from Komsomol, the youth organization of the Communist Party. To speed up the building process, the Central Committee launched a recruitment campaign among young communists (*komsomoltsy*), who turned out in great numbers. Driven by peer pressure, fear or in many cases what was probably a sincere patriotic spirit, some 500 *komsomoltsy* filled the train coaches in Leningrad and headed for the makeshift platform in Zapolyarnyi, where they were warmly greeted by the first settlers. The Soviet romanticism that permeated the Komsomol discourse is clearly demonstrated in the salutation presented by fellow youth communist Lyusya Vasileva upon the arrival of one of the trainloads to Murmansk, en route to Zapolyarnyi: 'We truly and honestly, friends, like it here. . . . Together we will build houses and start in this place a wonderful, big and friendly Komsomol family.'[3] And, in a way, that is exactly what they did. A mere two years after the foundation stone was laid down in May 1955, Zapolyarnyi was declared a full-fledged proletarian village (*rabochii poselok*) of the Soviet Union. The industrial activity in and around the Zhdanovskii mines and the adjacent enrichment plant gave room for even more growth, and in 1963 Zapolyarnyi's status was upgraded from village to town (*gorod*).[4] Zapolyarnyi thus became the first town in Pechenga – Nikel still counted as a village.[5] By 1964, Zapolyarnyi had a population of 12,000.[6]

The arrival of the first Nikel-bound train in 1956 was another major advance in the history of the Pechenga region. Between 1951 and 1953, a separate correctional labour camp (*ispravitelno-trudovoi lager*) had been set up for the purpose of laying the railroad tracks from Kitsa, situated south of Murmansk, all the way to Nikel, although the northernmost stretch from the river Zapadnaya Litsa was left for others to build.[7] The completion of the railroad to Nikel was a welcome event. It allowed the Komsomol enthusiasts in Zapolyarnyi to travel by train all the way to their destination and provided the logistical framework needed for the settlement of what was basically a semi-wilderness. It was also equally important as an eagerly awaited alternative transport route between Pechenganikel and the rest of the Soviet Union – both for transport of nickel southeastwards and for supplies to reach Pechenga. Furthermore, since railroad tracks were also laid to the Pechenga fjord, communication between the ports there (Pechenga and Liinahamari) and Nikel became more reliable than the erratic logistics experienced during the early post-war period. Thus, the introduction of rail communications to remote Nikel was an important step towards the further incorporation of the combine into the Soviet economic realm, which had been pursued with intensity from the very moment of Moscow's acquisition of the area in 1944.[8]

Pechenganikel raised its productivity, and this was not only due to the newly discovered ore deposits. The original mine Kaula, which had been the basis for Finnish–Canadian industry in the 1930s, was still yielding immense amounts of nickel ore. The miners dug deep underground, and in 1958 reached the 'eighth floor', counting from the top downwards. In 1961, the neighbouring deposits at Kotselvaara were opened for production, and the main mine in Nikel was from then on called Kaula-Kotselvaara.[9] Also, the Kammikivi deposits, which had been located by the Finnish geological commission in the 1930s,[10] were further examined and deemed worthy of exploitation. Production commenced in 1953. Combined with the Kotselvaara addition, this provided the basis for another enrichment plant in Nikel, which was active from 1958. Advances in ore refining were also important in raising productivity. Between 1962 and 1967 new electric furnaces and convertors were installed in the smelting plant in Nikel, and enhanced experimental capacity allowed for a higher degree of utilization. In 1965, an enrichment plant in Zapolyarnyi started producing concentrate; and, in 1971, a new wing of the Zhdanovskii deposits near Zapolyarnyi, the Vostok (eastern) mine, yielded its first ore. In 1975, the Severnyi (northern) deposits were opened for production.[11] By the mid-1970s, the once-demolished war booty had become a highly productive Soviet industrial area.

A parallel development, and logical consequence of the expansion of exploitable ore deposits in Pechenga, involved the almost continual efforts to expand the base for electric power supply from the Pasvik River rapids. Intensified industrial processes and a growing population in Pechenga contributed to an explosive increase in power consumption in the Pasvik valley. From 1947 onwards, the Finnish firm Imatran Voima partly rebuilt, partly constructed three power plants on the upper stretches of the Pasvik River for its Soviet customer. The reconstruction of the Jäniskoski power plant was completed in 1951, and the rapids at Rajakoski (1956) and Kaitakoski (1959) were also tamed to provide energy to the Soviet settlements in Pechenga. These three power plants were all located on what was uncontested (after the Soviet purchase of the Jäniskoski area in 1947)[12] Soviet territory in the upper 30 km of the Pasvik River.

Nonetheless, power consumption quickly outpaced production at the three plants, and Soviet authorities were soon looking for new energy sources. The lower 110 km of the Pasvik River, which were shared with Norway except for the Borisoglebskii enclave, contained several whitewater sections that were suitable for hydropower development. To realize this potential, however, Soviet and Norwegian authorities would have to resolve various issues related to the two countries' rights to utilize the river.

When the question was brought up in the mid-1950s, it was not an unheard-of topic in Norwegian–Soviet relations. Already in July 1945, when the victorious Soviet Union enjoyed an unprecedented degree of good will in Norway after having liberated the country's northern parts from the Nazi forces, a Soviet initiative had been taken. In a note to the Norwegian embassy to Moscow, it was stated that Pechenganikel's future power needs were to be satisfied by the building of a power plant in Borisoglebskii.[13] The Norwegian reaction was hesitant, and further notes were exchanged over the next two years.

Several factors would put Soviet plans to rest for the time being. On the Norwegian side, willingness to enter bilateral collaboration on the border with the Soviet Union diminished as world politics headed towards a sharper East–West divide. Thus, Norwegian security concerns over time trumped the economic gains expected to accrue from the country's rights to one-third of the power generated by the prospective Soviet power plant.[14] More important, though, was the development in the parallel Soviet–Finnish talks concerning the Jäniskoski territory and power station.[15] The negotiations resulted in full Soviet access to the upper stretches of the Pasvik River, rendering Moscow's ambitions for Norwegian–Soviet collaboration on hydropower less pressing, for the time being at least. The Borisoglebskii plans were relegated to the back burner.

Soviet overtures relating to hydropower on the Pasvik River re-emerged in the somewhat milder East–West climate following Stalin's death in 1953. During negotiations in Oslo in November 1955 on a related matter,[16] the possibility of a Norwegian–Soviet power plant project was again mentioned by Soviet representatives. This time, although Norway had signed the Atlantic Pact in 1949 and the Pasvik River thus in effect marked the frontier between NATO and the Soviet Union, the post-Stalin thaw had created a political atmosphere more conducive to realization of such plans. Originally, the Soviet negotiators had envisaged two joint power plants that would deliver energy to both Norwegian and Soviet consumers. This was emphatically opposed by the Norwegian side, and the parties agreed to build separate power plants. However, in a parallel to the Jäniskoski arrangement, the Soviet side insisted that the power plant in Borisoglebskii be built by a Norwegian company. After some hesitation, Norway accepted this.[17] In 1960, agreement was reached between the Norwegian company Norelectro and the Soviet Ministry of Power Stations (Minelektrostantsii), and construction of the Borisoglebskii power plant could commence.

The prospect of a Norwegian company building a power plant on the Soviet side of the national border was not a straightforward matter in the late 1950s. Although it would have a positive effect on local employment rates in Finnmark, the Norwegian security establishment opposed this solution, fearing Soviet attempts to recruit agents among Norwegian workers. These worries were to some extent justified: the Finnish security police had informed their Norwegian colleagues of several instances where Soviet intelligence personnel had recruited from Imatran Voima's employees at Jäniskoski. It was feared that the Borisoglebskii project would give ample opportunity for Soviet scouts to establish contacts on the building site – which did take place, to some extent.[18] The Norwegian secret police, having failed to stop the project, were probably relieved when the Borisoglebskii power station was completed in 1963.

However, the Norwegian need to control traffic in the border area was challenged once again a mere two years later. When in the summer of 1965 Soviet authorities surprisingly opened up the Borisoglebskii enclave for Norwegian tourists, the Norwegian security establishment was even more wary. Again, Soviet agent recruitment was the main problem. The Norwegian authorities closed the border for eastward traffic only three months after the area had been opened by the Soviet Union. Norwegian fears of Soviet intelligence gathering were probably somewhat overblown. Although the presence of known Soviet agents in the enclave was noted by Norwegian observers, there is little evidence to suggest that espionage was particularly widespread. The confined and easily surveyed Borisoglebskii area was hardly an ideal arena for conspiratorial activity. Rather than being an element in a Soviet espionage scheme, the opening of the Borisoglebskii border can be seen as part of an overarching propaganda push directed towards Western states, including Norway. Demonstrating peaceful intentions and a will to engage with Western countries was a central message in Khrushchev's period at the helm in Moscow.[19]

This brings us to the question of why the Soviet negotiators were so eager to have a Norwegian company construct the Borisoglebskii power plant, rather than doing it themselves. Originally, as mentioned, they had even suggested that Norway and the Soviet Union build joint power plants. Various factors may have led to these rather remarkable Soviet positions. First of all, Norway's expertise within the field of hydropower was well developed, so the opportunity to gain insight into Norwegian technological solutions may have been a point of interest. Secondly, as mentioned, the possibility of recruiting Norwegians to work for Soviet intelligence was obvious, although it was probably a secondary concern. Three other explanatory factors are more convincing. Soviet expertise did exist but was in short supply. There was simply a lack of qualified personnel to do the job. Using first Finnish and later Norwegian workers and engineers in the border valley freed up much-needed Soviet personnel for other projects. Also, the fact that the Pasvik projects were located right at the border probably made the use of a large number of Soviet workers there unattractive to a Soviet leadership that by sheer reflex would aim to keep the population out of touch with the outside world. Thirdly, the Soviet strategy of presenting the USSR as peace-loving and willing to cooperate with the Western world was at this time the central message in Soviet self-presentation to the world, albeit hardly a new one.[20] Extensive cooperation with NATO-member Norway would, in the eyes of Moscow, very publicly demonstrate the Soviet striving for peaceful relations with ideological opponents.[21] From a propaganda viewpoint, construction of the Borisoglebskii power station was thus seen as very useful.[22]

By 1963 the Borisoglebskii power plant was generating electricity.[23] Yet another three power plants would be built on the Pasvik River, of which one was located on Russian soil. The two Norwegian power plants at Skogfoss and Melkefoss were completed in 1964 and 1978, respectively. Hevoskoski power plant, a fully Soviet venture, was started in 1956. Its construction period was the longest of all for the power plants in the Pasvik valley – fourteen years. Hevoskoski generated electricity from 1970. In the course of the first two decades after the Second World War, then, the Pasvik River was thoroughly regulated. Seven power plants were built, only one of them by Soviet entrepreneurs. Nevertheless, five of the power plants were Soviet property – three of them built by Finnish Imatran Voima and one by the Norwegian company Norelectro. Two of the Pasvik power stations were Norwegian.

In sum, industrial progress in post-war Pechenga was immense. This expansion was felt also outside the production cycle. Improvements were made in the everyday lives of the residents of Nikel and Zapolyarnyi, and the hardships experienced when the area first became Soviet property abated as various facilities became accessible.[24] By the mid-1980s, the region boasted two health clinics, two sports complexes with indoor swimming pools, two 'houses of culture', four schools, fifteen nurseries, a department of the Monchegorsk Technical College and an outdoor recreational zone. Meat became increasingly available in 1984 with the establishment of a hog farm with annual output capacity of 30 tons of pork. Furthermore, all industrial facilities were equipped with canteens, and residents were regularly given free or low-cost vacations to southern parts of the USSR. Perhaps most important, average living space per inhabitant was more than double what it was in 1951, when it had been a meagre 4.69. In 1984, the average *zapolyarnets* and *nikelchanin* enjoyed 10 square metres all to himself or herself.[25] However, all these material improvements were accompanied by undesirable side effects. The flip side of the coin was indeed problematic: environmental degradation was becoming extensive in the Pechenga region.

## Pollution in a socialist state

Well into the 1970s, it was widely assumed in Western societies that industrial pollution, or environmental disruption caused by production processes, was a phenomenon mostly restricted to capitalist economies. Experiencing the difficulties originating from emissions to water and air, Western academics – especially those who in the 1970s were fashionably influenced by Marxist ideology – would present the problem of environmental disruption as one inherent to the capitalist organization of production.[26] The reasoning was that private industrialists would, in the absence of rigid state regulations, take little interest in making large investments in order to prevent environmental damages stemming from their production cycle. Once the emissions had left their factories, either via smokestacks or through pipes leading wastewater to nearby streams, pollutants were no longer the individual industrialist's responsibility. They became a societal problem that would have to be dealt with by public agencies.

This critique of private industrial owners' lack of environmental consciousness was both precise and easily documented in most Western countries. One logical, albeit very hasty, conclusion to this line of thought was that in socialist countries, where industries were state-owned and the producer (the state) was responsible not only for

Worker in the Pechenganikel smelting plant wearing protective inhalator.
Photo: Ola Solvang

the profitability of industrial production but also for the welfare of the
population, environmental disruption would hardly occur. Studies
of Soviet law gave support to the assumption that Soviet enterprises
were subject to stringent pollution control and that the socialist
state had – as any rational socialist state presumably would – done
away with environmental disruption through legal measures. These
misinterpretations became taken for granted in Western debates about
industrial pollution due to a lack of solid environmental data from
behind the Iron Curtain. Only with the gradual opening up of Soviet
society to Western observers would the environmental calamity of
communism become more apparent.[27]

The myth of pollution-free socialist countries was, not surprisingly,
embraced in Soviet rhetoric. The Soviet legal code on pollution control
was restrictive and would be cited as proof of the socialist system's
ability to eliminate environmental disruption.[28] The mere existence of
laws, however, is no guarantee of prohibitive effect. The maverick Soviet
ministry official Zeev Wolfson, writing in the late 1970s under the
pseudonym Boris Komarov, amply showed how Soviet environmental

legislation was rendered impotent in a federal system that was first and foremost preoccupied with material advances and industrial development.[29]

Although the scope of environmental destruction was unknown to the population at large, pollution was always a palpable part of everyday life in Soviet industrial towns. Already in the late 1940s and early 1950s, the Pechenganikel management reported that waste gases were pouring out of the smelters, resulting in badly deteriorated air quality in and around the industrial site. At the time, the concern was mainly for the negative effects it might have on human health, not the damage to the natural environment. An account from a trip made to Nikel in 1960 focused more on the damages to Pechenganikel's surroundings. Grigorii Svirskii, a correspondent for the Soviet weekly *Ogonek*, was appalled by what he saw:

> The closer I got to the factory, the stronger a feeling of wariness and vexation gripped me. An enormous smokestack towers over the Pechenganikel combine. It is the second tallest in the Soviet Union, I later found out. The grayish column of smoke from the stack is taken far away. But for some reason everything around the factory – the squat arctic birches and the sparse pine trees – has all gone black and shriveled from sulfurous gases. Having had a closer look, I see the gases whirling just above ground, poisoning everything living. . . . What is this? An accident? Or is this how they deal with waste gas *[zagazovannost]* in the workshop here?[30]

Interestingly, Svirskii did not seem to worry about the column of smoke rising from the smokestack, as it was 'taken far away'. Indeed, tall smokestacks were built as a solution to local environmental problems, as they would discharge waste gases in a layer of the atmosphere where they could be absorbed and dissolved. Such absorption and dissolution, however, presupposes that the discharges are kept at a fairly low level.[31] When Pechenganikel started importing ore from the nickel mines in Norilsk in the 1970s, a new situation emerged. The Norilsk ore was, unlike that in Pechenga, highly concentrated and could therefore be processed without preliminary enrichment. In environmental terms, though, the Siberian ore was more problematic. It had a very high sulphur content, which meant that discharges of sulphur dioxide from the Pechenganikel smelting plant increased markedly in the 1970s.[32] From having annually discharged approximately 100,000 tons of sulphur dioxide before the Norilsk ore was introduced, a massive discharge of 400,000 tons was measured in 1979.[33] Very soon, the effects of the Norilsk ore imported to Pechenganikel's smelters were felt across the borders.

When Anders Aune, the governor (*fylkesmann*) of the northernmost Norwegian county Finnmark, in 1978 approached the Murmansk oblast administration, complaining about how sulphurous discharges from Pechenganikel had negative effects on the natural environment on the Norwegian side of the border, he was immediately appeased. Murmansk officials were quick to explain that there were Soviet plans for a full reconstruction of the nickel plant, and that emissions would be back at the more normal level of 100,000 tons annually by the end of the year.[34] However, the announced improvement did not occur, and we may see this first exchange between the county of Finnmark and Murmansk oblast as the early start of what was to become a long and troublesome multinational effort to reduce emissions from Pechenganikel. As we shall see in the following section, the late 1970s gave reason for optimism in the Nordic countries with regards to Soviet environmental ambitions – an optimism that must be understood in the broader context of international environmental cooperation.

## Environmental détente

True to its traditionally wary approach to multilateral collaboration with Western capitalist countries, the Soviet Union was no leading force when environmental protection entered the international agenda in the early 1970s. Nevertheless, the decade of 'the first green wave' in international politics,[35] which was also the decade of East–West détente, did produce some remarkable changes in this respect. Faced with increasing pressure on the Soviet economy, the Kremlin leadership was obliged to ease international tensions – both to be able to reallocate resources from the very costly arms race with the United States to domestic production and to become more palatable to Western creditors. General Secretary Leonid Brezhnev was instrumental in starting a new chapter in Soviet foreign policy that was to be characterized by a more open approach to international cooperative efforts. The combination of the resulting détente, which led to several historic East–West agreements,[36] and the simultaneous emphasis on environmental matters in international politics, prompted an unexpected 'greening' of Soviet approaches to international relations.

One prominent feature of this 'greening' – not without relevance to the nickel industry on the Kola Peninsula – was the Soviet Union's pivotal role in the inception and subsequent signing of the Convention on Long-Range Transboundary Air Pollution (LRTAP) in 1979. Sweden first brought the matter of transboundary air pollution to international attention. In 1968, Swedish researcher Svante Odén established how sulphurous gases emitted in Britain and Central Europe were transported by prevailing northeasterly winds before falling as acid precipitation over Nordic soil and waters. His findings were, naturally enough, taken very seriously in neighbouring Norway, but did not arouse much concern on the British Isles or among the polluting states in continental Europe.

In their push for international recognition of this environmental problem, Sweden and Norway found an unlikely ally in the Soviet Union. From 1975 onwards, Leonid Brezhnev promoted the issue as part of his campaign to preserve détente and even enthusiastically supported the Nordic countries in their 1977 call for binding international commitments to reduce emissions of sulphurous gases.[37] It was hardly environmental awareness on the part of Brezhnev and his party colleagues that brought this about. To the Kremlin, the issue of transboundary air pollution was appropriate for demonstrating a degree of 'cooperativeness', which was important if targets in more important areas, mainly trade with Western states, were to be achieved. Environmental cooperation, unlike human rights issues, did not threaten the internal stability in the Soviet Union and seemed ideal for the purpose of exuding willingness to interact constructively with Western countries.[38] Soviet support became even more pronounced when Gro Harlem Brundtland, at the time Norwegian minister of environmental affairs, brought up the acid rain issue during a visit to Moscow in 1978. One result of this visit was collaboration between Norwegian and Soviet bureaucrats to promote an international regime for monitoring and reducing transboundary air pollution.[39]

When the LRTAP Convention was signed in Geneva by thirty countries in November 1979, the way had been paved by Scandinavian diplomacy, supported by the Soviets. The main proponents of LRTAP, Norway and Sweden, were pursuing an offensive strategy to make the new regime effective. They promoted 30 per cent cuts in sulphur emissions across the board in Europe. Faced with this radical demand, the Soviet leadership became, albeit haltingly at first, convinced that a diplomatic victory at the expense of the United States could be won at low cost. In 1985, when the first sulphur protocol under the LRTAP regime was signed in Helsinki, the Soviet Union was one of nineteen signatory states.

In fact, there were various mitigating factors that made the required emission cuts easy to handle for Soviet industry. Firstly, Soviet negotiators insisted that the cuts should be applied only to emissions that crossed the western borders of the USSR. The major part of Soviet emissions fell over the Soviet Union itself and was therefore exempted from the first sulphur protocol. Secondly, in the balance sheet of transboundary air pollution, the Soviet Union was (like the Scandinavian countries) a net importer of acid precipitation, due to the prevailing wind patterns. This in itself secured Soviet compliance with the LRTAP regime. Thirdly, and most importantly, sulphur emissions along the Western borders had already been curtailed as a result of domestic Soviet policies completely unrelated to the required emission cuts in the sulphur protocol: due to a dramatic decline in oil and coal production in the European parts of the USSR from the mid-1970s onwards, the Soviet leadership had replaced, or was in the process of replacing, these highly sulphurous fuels with nuclear energy and natural gas. This transition alone accounted for most of the 30 per cent cuts stipulated by the sulphur protocol.[40]

Thus, Soviet compliance with the LRTAP Convention did not necessitate any far-reaching structural changes that had not already been set in motion for other reasons. Any increase in Soviet international credibility accruing from this demonstration of 'cooperativeness' therefore came cheaply.[41] As we shall see in the coming chapters, the lax Soviet approach to the LRTAP regime was evident also in the case of the Kola Peninsula nickel industry and most prominently at the Pechenganikel combine. Here, the intentions of the LRTAP Convention, although highly applicable, were mostly ignored, as the Soviet authorities failed to implement domestic measures that would decrease sulphurous emissions. Implementation was deemed unnecessary, as exogenous factors, like the transition from carbon fuels to natural gas and nuclear power, had brought sulphur dioxide levels sufficiently down to be in compliance with LRTAP targets. On the other hand, there is reason to believe that the LRTAP commitment did bring about some behavioural changes in the Soviet bureaucracy and research milieus that were exposed to new interfaces with Western colleagues, and who were working earnestly to diminish airborne pollution.[42]

Political scientist Robert G. Darst makes a compelling and well-founded case for how the Soviet approach to the LRTAP regime, as well as to subsequent protocols under its auspices, was motivated by anything but environmental concerns. He claims it was a defensive strategy at a time when the Soviet leadership wanted to divert international attention from 'hard policy' areas such as defence to 'soft policy' areas such as environmental issues. By creating opportunities for East–West cooperation in these areas, Soviet diplomacy could ease international tension, and thereby relieve the strain put on the Soviet economy by the spiralling arms race. In fact, it was a matter of sustaining the faltering Soviet economy and, in consequence, of Soviet survival. In its struggles to preserve supremacy in Eastern Europe, argues Darst, the Soviet leadership portrayed itself as 'cooperative' through involvement in multilateral collaborative structures that would demand little effort, if any effort at all. As Darst and others have shown, Soviet costs associated with the LRTAP regime were miniscule – as were the actual environmental effects within the Soviet Union stemming from participation in the regime.[43] It is easy to concur with Darst and others in their verdict that Soviet environmental commitments in the Brezhnev era were made with a view to 'high politics'.[44] It seems very likely that what motivated the Soviet leadership was not protection of the environment, but the need to gain ground in other areas targeted in Soviet foreign policy – like preserving détente.

This phenomenon was not restricted to the Brezhnev era. Both Gorbachev's willingness to address problems of industrial pollution and subsequent Russian environmental policies under Boris Yeltsin and Vladimir Putin were arguably shaped by similar hidden agendas. In the case of Gorbachev, Darst is in no doubt when comparing him to his predecessor:

> Gorbachev, by contrast [to Brezhnev], was committed to ending the Cold War altogether and to the integration of the USSR into the world capitalist economy. He was therefore willing to sanction a massive redirection of Soviet resources toward the environmental issues of greatest interest to the West – an orientation that generated the unprecedented 'greening' of Soviet foreign policy in the second half of the 1980s.[45]

In other words, Darst sees Gorbachev's environmentalism as merely a means by which other and more important goals could be pursued. In much the same vein, Putin's (and, in practice, the Russian State Duma's) decision to ratify the Kyoto Protocol has been seen as motivated by the expected enhancement of Russia's international image and economic gains from carbon trading.[46] Before turning to post-Soviet developments, however, let us take a closer look at environmental policies during the Gorbachev's period at the helm.

## Environmental perestroika

As we have seen, Pechenganikel entered a long period of industrial development after Imatran Voima's construction projects in the upper Pasvik valley were concluded. Although the adjacent borderline with Norway experienced some international activity, such as the Norwegian power plant construction and the brief border opening for tourists in Borisoglebskii, the Pechenganikel combine itself mostly escaped the Western gaze. This interim period of relative calm was to end abruptly in the mid-1980s. No longer was it the industrial expansion of Pechenga, but rather the destructive potential of the industry placed there, that attracted attention. And this development was made possible by radical changes in Soviet approaches to both domestic and international politics.

For a while, Mikhail Gorbachev's introduction of openness to Soviet political life amounted to merely launching a slogan: glasnost. However, the events of April 1986, when a massive environmental disaster came to loom over western parts of Soviet territory, clearly demonstrated the need for openness and the dangers of traditional Soviet secrecy.[47] Although the Chernobyl disaster did not elicit anything amounting to the full riot that would later bring the Soviet Union to its end, it did ignite a flame that would put environmental concerns at the heart of popular opposition to Soviet authorities. In the following years, environmental issues would fuel dissatisfied voices in the increasingly bold Soviet public sphere. Environmental protection became a central part of the politics of glasnost and perestroika, at least initially. Although hardly rife with environmental activism, the mono-industrial company towns of Murmansk oblast would also be affected.

A Soviet brand of environmental activism already existed by the 1980s. In fact, it had long traditions, having survived both the early revolutionary years as well as the subsequent upheavals of Stalin's campaigns to collectivize and industrialize Soviet society. Contrary to what one might expect, the movement's critical agenda was tolerated by the Soviet leadership.[48] A possible explanation is simply that the early Soviet environmentalists were hardly a threat to industrial interests

View of Nikel from the southwest.
Photo: Ola Solvang

but championed primarily conservationist objectives instead. This included the protection of endangered species and cultural heritage and, most importantly, preservation of pristine nature through the establishment of a series of nature protection areas (*zapovedniki*).[49] Nature conservation did not interfere significantly with the overriding ambitions of the Soviet state – natural resources of industrial significance were still readily available to the main actors of the Soviet economy. Thus, the movement could become part of the flora of semi-official organizations condoned by the authorities.[50]

Although highly critical of Soviet industrial progressivism, this traditional movement was not able to keep up with the societal transformation of glasnost. Despite the dramatic lessening of authoritarian constraint on grassroots organizations in the latter half of the 1980s, it could not make the leap from conservationist objectives to the more pressing issues of industrial pollution and its detrimental effects on people's health and daily lives. New groups arose, focusing on the immediate needs of Soviet citizens to breathe fresh air and drink pure water. It was they who took the lead in the glasnost-induced opposition.

By 1987, ecological problems had become the foremost single issue in public criticism of the Soviet authorities. The essence of this opposition, however, was not a deeply rooted concern for the environment. Instead, these democracy activists were demonstratively and euphorically exercising the right provided by glasnost to point to governmental deficiencies, and environmental degradation was a topic that could safely be taken up. Thus, anti-pollution campaigns briefly served as vehicles for anti-Soviet sentiments, but the environmental activism soon gave way to other concerns as a progressively emboldened Soviet public moved on to the core questions of their existence. Historian Douglas R. Weiner has described this development very accurately:

> By the early 1990s, as purely economic and political issues edged out even the urgent concerns about public health, as workers were now forced to choose between slow poisoning and unemployment, the fight against pollution did not seem nearly as clear-cut an issue as it had a mere three or four years earlier. For many, putting bread on the table is more urgent than shutting down a factory that causes asthma in a child. Both usually take precedence over fighting for a national park.[51]

On the Kola Peninsula, the subordination of environmental concerns to issues such as employment and material survival was even more pronounced than in more central parts of the Soviet Union. In fact, Murmansk oblast experienced neither much environmental activism nor political activism in general. The region was economically controlled by several large industrial enterprises, such as Pechenganikel, and their ministerial superstructures. While these polluting enterprises were responsible for extreme environmental degradation of the most densely populated areas, they also commanded the loyalty of their employees. These workers would hardly rally against the cornerstone factory which provided them with their livelihood, as illustrated in the quote provided in the previous paragraph. The lack of oppositional activity in Murmansk oblast was compounded by the unusually strong political control, as the KGB and the Communist Party enforced a particularly strict regime in the highly militarized border region.[52] Thus, glasnost simply did not provoke the same reactions in the streets of the Kola Peninsula company towns as those seen in other areas of the Soviet Union with more diversified economies.

That said, despite the absence of an activist population, environmental issues were raised. Rather than as part of an environmental campaign from below, pollution problems in the Russian northwest were addressed from the very top. When Mikhail Gorbachev visited Murmansk on the first day of October 1987 to declare the Northern capital a 'hero city' (*gorod-geroi*), he made a speech that has been considered pivotal in the process that was to lead towards the many ensuing collaborative efforts between the Nordic countries and the Soviet Union (and later the Russian Federation).[53] Speaking about the environment, the radical general secretary was as clear as anyone could have hoped for:

> We attach special significance to the [future] collaboration between northern countries within the sphere of environmental protection. The urgent necessity of this is obvious. . . . The Soviet Union proposes a concerted development of a joint comprehensive plan for the protection of northern nature. Northern European countries could stand as an example to others in agreeing to the establishment of control systems for the state of the environment and for radioactive security in the region. It is necessary to make haste in the work to protect the tundra and the taiga areas of the north.[54]

Gorbachev's words were in reality as much a commentary on already existing processes as a declaration of future plans. To conclude that this inviting speech gave the impetus to, for example, the environmental collaboration between the Soviet Union and Norway would be to overestimate the general secretary's rhetorical abilities.

In fact, discussions between Norway and the Soviet Union on environmental matters had been going on since before Gorbachev came to power, although they gained pace only under his leadership. Already in 1984, the county governor of Finnmark had approached the Murmansk administration with a view to bilateral talks about the environmental situation in the border areas. A first regional meeting took place in June 1986, as parallel talks on ministerial level were nearing their conclusion: a Soviet–Norwegian environmental agreement was planned for signing in July 1986. The Chernobyl accident in April that same year, however, prompted the Norwegian minister of the environment to require that radioactive pollution be included as one of the topics in the agreement, which led to a postponement.[55] Also the Soviet side had last-minute additions to the text, which further delayed the signing. Eventually, the two states reached an agreement on the content of a collaborative environmental programme. The final details were agreed during the spring and summer 1987 – well before Gorbachev gave his now-famous speech in Murmansk in October.[56] Some time would pass, however, before the agreement was signed by representatives of the two states.

Although motivated by the need for advances in other foreign policy areas, the 'greening' of Soviet foreign policy from the Brezhnev era through to the Gorbachev period did influence groups within the Soviet bureaucracy that were involved in environmental cooperation with the West. In the final decade of Soviet existence, these groups, such as the State Committee for Hydrometeorology and Environmental Monitoring (see Chapter 3), were able to campaign successfully against more powerful agencies such as the KGB and the industrial ministries to be able to share Soviet environmental monitoring data openly with Western colleagues. Environmental cooperation gradually entered a more constructive phase. Although total Soviet sulphur dioxide emissions were within the LRTAP targets, Pechenganikel was still subjected to demands for discharge reductions. In 1980, the Soviet government even agreed to cut emissions that were transported into neighbouring Finland by 50 per cent in 1995, which was more than required by the LRTAP sulphur protocol.[57] However, although the LRTAP target of 30 per cent cuts from 1980 level was reportedly reached at Pechenganikel in 1992,[58] the emissions were still considered unacceptable in neighbouring Norway and Finland.

*       *       *

As the Soviet state was drawing its last breath, Pechenganikel would find itself at the centre of international attention. In the years leading up to the Soviet collapse, it was the airborne pollution from the combine's smelters that would put Pechenganikel in the international spotlight. These industrial discharges provided the core motivation for Norway's initiative to establish a joint Norwegian–Soviet Environmental Commission.

In the following chapters we examine Soviet and later Russian environmental politics in the Gorbachev, Yeltsin and Putin eras. The emphasis is not on the overarching policy making as such. Issues will be indirectly illuminated through a closer look at how the problems of pollution from the Pechenganikel combine were handled.

In the aftermath of the Soviet breakdown in 1991, a completely new political reality was to emerge. The Boris Yeltsin decade – the 1990s – was characterized by a strong diffusion of power from the Moscow centre to regional authorities, and this was also evident in environmental politics and practices. Initially, this empowerment did seem to influence regional policies on environmental affairs. The Murmansk oblast administration declared its firm interest in improving conditions in the company towns around the Kola Peninsula. In hindsight, and with special regard to Pechenganikel, we can see that these ambitions were not realized.

We will in the following chapter discuss why there was never an environmental breakthrough in this period. Later, a new regime ushered in a period of fresh hope in Russia. When Vladimir Putin came to power at the turn of the millennium, he was to reverse many of the processes that had taken place in Yeltsin's decentralized federation and re-established the traditional vertical power and its apex in the Kremlin. It soon became clear, however, that Putin's 'managed democracy' could not provide an answer to the Soviet environmental dilemma. The reasons for this will also be illuminated in subsequent chapters, through a close look at developments in Pechenga.

# Business, environmentalism
# and the Soviet collapse

Glasnost was more than just an internal Soviet phenomenon. As a result of Gorbachev's new political strategy, the outside world gained insight into matters previously unknown to all except those within closed circles. This was abundantly true within the field of environmental pollution. Gradually it became clear that more than sixty years of intensive Soviet industrialization and breakneck utilization of natural resources had left their mark on the environment in and around many densely populated areas. In the late 1980s, 'the Soviet ecocide' became a topic in Western countries, especially those that shared borders with the Soviet Union and were directly affected by pollutants from its industry.[1] To the Nordic countries, the sulphur dioxide emissions from Nikel and Monchegorsk were of particular relevance, and newly available information about their extent and negative effects gave cause for growing concern.

This chapter covers the initial phase of Nordic cooperation with the Soviet state to curtail sulphur dioxide emissions in the border regions and introduces various issues that formed the backdrop to the negotiations. We begin by discussing the diverging starting points for Nordic and Soviet civil servants that became apparent in their early meetings about environmental protection. These differences were also reflected in national understandings of environmental issues, specifically the sulphur dioxide pollution from the nickel works, as becomes apparent in a review of Norwegian popular engagement and Soviet reactions to this. We next examine the non-environmental interests at play in the cooperation around the nickel industry.

While Soviet benefit seeking in environmental cooperation is widely acknowledged, the commercial aspirations of the Nordic countries are also an important, but often overlooked, factor. In this chapter, we see that environmental as well as industrial and commercial concerns abounded. On both sides of the negotiation table, there were ambitions that went far beyond just solving the problem at hand. A comprehensive solution to the industrial pollution at Pechenganikel, involving a massive reconstruction of the industrial installations, provided commercial and political perspectives that were to have much more far-ranging consequences than simply dealing with the damaging sulphur dioxide and heavy metals. The emissions from Nikel became more than just a problem to be solved – they became a potential political and economic resource as well.

## An odd couple

The Soviet–Norwegian Environmental Agreement, establishing the joint Soviet–Norwegian Environmental Commission, was signed in Oslo on 15 January 1988, by Prime Ministers Nikolai Ryzhkov and Gro Harlem Brundtland.[2] As we saw in the previous chapter, government officials in Moscow and Oslo and regional authorities in Finnmark and Murmansk had for several years been holding talks about environmental issues. With the establishment of the environmental commission, these issues could be discussed in a more permanent bilateral setting. At the centre of attention were the sulphur emissions from Pechenganikel – emissions that were deemed undesirable by both parties. However, as we shall see, although both Soviet and Norwegian actors agreed that environmental issues were important, the two parties approached the matter at hand very differently. The commission joined two countries of very dissimilar political, cultural and scientific leaning and traditions, as was reflected in the proceedings. Even more important was the chasm in world views between the citizens of an affluent small state like Norway and those of a faltering and poverty-stricken socialist giant like the Soviet Union. It is likely that this disparity was at times revealed in very different goals and intentions, some of them unspoken.

Nikel's location less than 10 kilometres east of Finnmark, the northernmost county of Norway, made discharges from the Pechenganikel combine palpable also in Norway. The smell of sulphur in the air was unmistakable; even more alarmingly, the branches and leaves of polar birches in Eastern Finnmark were periodically scorched by sulphurous air streams, and reindeer fences in the border area corroded with remarkable speed. These observed damages were immediately attributed to the Pechenganikel emissions. Indisputable documentation was required, however, before formal complaints could be lodged. In 1974, the Norwegian Institute for Air Research (NILU) started systematic measurement of air pollutants in the border area. The measurements made over the following ten years enabled the Norwegian side in 1986 to conclude that the Pechenganikel emissions were the main cause of harmful effects observed in Finnmark. These findings were presented to Soviet representatives at the first meeting on environmental issues between Finnmark County and Murmansk oblast in 1986.[3]

Although conducted in a most amicable atmosphere, the initial meetings between Norwegian and Soviet representatives were indicative of a divide in the parties' approach to nature management. True, in his opening statement to the regional meeting in Kirkenes in June 1986, first deputy chairman of the Executive Committee of Murmansk oblast Sergei Zhdanov distanced himself, and his compatriots, from the traditional Soviet perception of the natural environment: '[The] understanding that until recently was commonly held, i.e. the storming of nature, the fight against nature and the confrontation between nature and human enterprise, is now history.'[4] This statement, however, merely shows Zhdanov's ability to bend realities when meeting a group of environmentally minded Western neighbours. As this chapter will show, the traditional heavy-handed Soviet-style approach to the natural environment was still dominant.

Despite growing environmental concerns among the Soviet public by the early 1980s, pollution control and nature protection were not yet prominent policy areas. In 1986, there was still no designated environmental ministry in the Soviet Union. Environmental issues, to the extent they were at all deemed pertinent, were briefly addressed by the sector ministries in their five-year production plans. Any measures taken would be communicated directly from the ministry in Moscow to the subordinated enterprises. Regional authorities had no say in environmental matters.[5] The only non-industrial agency that could influence environmental decisions was the State Committee for Hydrometeorology and Environmental Monitoring (Goskomgidromet). It was involved in international air quality monitoring[6] and headed an interdepartmental commission appointed to oversee Soviet implementation of the LRTAP regime. Goskomgidromet, however, had little clout when confronted with industrial agencies like Mintsvetmet.

These Soviet realities were not mirrored in Scandinavia. In Norway, as in much of the Western world, environmental management had become an integral part of the political system with the first green wave of the late 1960s and early 1970s. When Norway established its Ministry of the Environment in 1972, it was seen as a counterweight to industrial interests, in the private and public sectors alike, and was equipped with the legal tools and the economic resources necessary for enforcement of environmental legislation.[7] In 1982, environmental divisions were established at all County Governor offices, thereby strengthening the regional enforcement of environmental laws and regulations.

The leading environmental light in Scandinavian politics at the time was the Norwegian prime minister Gro Harlem Brundtland. Having chaired the UN commission that in 1987 launched what has been labelled 'in all probability the most important report on the environment ever to come out of the UN system', titled *Our Common Future*, she was *the* foremost advocate for 'sustainable development', catch-phrase number one of the greening of international politics.[8] Brundtland had, before she rose to the pinnacle of Norwegian politics, headed the NME. At the helm of this youthful institution from 1974 to 1979, she had gained general recognition in Norway for the importance of environmental protection, also in cases where such concerns collided with the wishes of industrial actors.[9] In other words, when Soviet representatives met with their Norwegian counterparts for the first commission meeting in Oslo in August 1988, they faced a political culture that, with some justification, promoted itself as environmentally progressive.

It is fair to say, then, that the first Soviet–Norwegian environmental meetings were encounters between two very different traditions. Although both parties could observe the same reality, that pollution was a major problem in the Soviet Union, there were very diverging ideas about what needed to be done and, even more importantly, how soon. A Norwegian environmental bureaucracy, accustomed to standing up to industrial interests, met a Soviet counterpart that by reflex considered industrial development the unrivalled priority of society. In the Soviet tradition, pollution was not a problem in and of itself, but more a symptom of what was at times euphemistically described as the result of 'insufficiently comprehensive exploitation of minerals and bio-resources'.[10] Paradoxically to Westerners, Soviet measures that had environmentally beneficial effects were not directed primarily towards curtailing pollution as such: they were designed to ameliorate the problems that the pollution was seen as a symptom of – namely problems of inefficiency in resource exploitation.[11] In contrast to Norwegian environmental agencies that acted as 'nature's watchdogs' pitted against industrial interests, Soviet bureaucrats dealing with environmental problems were expected to assist national enterprises in their efforts to optimize productivity and minimize industrial waste.[12]

The chasm between Soviet and Norwegian approaches to the environment is evident from the minutes from the regional meeting in 1986. While the Norwegian side talks about protection of endangered wolves, the Soviet side informs about its efforts to exterminate wolves to protect other wildlife. While the Norwegian side systematically presents proof of the continuing harmful effects of Pechenganikel on the environment in the border area, Soviet representatives give accounts of how well suited their legal framework is to combat pollution and extol the dramatic environmental improvements in Soviet industry. Interestingly, all Soviet speakers employ the expression 'efficient resource exploitation' when talking about environmental protection, reflecting an understanding of environmental issues as an integral part of the overriding industrial process.[13]

These differences were felt also at the ministerial level. In negotiations concerning the final text for the Soviet–Norwegian environmental agreement – the superstructure of the joint Soviet–Norwegian Environmental Commission – the Soviet side suggested the following passage: 'This cooperation aims to solve important issues of nature protection and *problems concerning the efficient exploitation of natural resources* [author's italics].'[14] In the final version, this sentence was changed, presumably according to Norwegian preferences: 'This cooperation aims to solve important issues of environmental protection and *to preserve the ecological balance* [author's italics].'[15] Thus, the end result was in line with the Norwegian inclination to see environmental objectives as worthy in and of themselves, rather than in accordance with the traditional Soviet perception of environmental problems as intertwined with, and subordinated to, industrial processes.

Soviet agreement to this essentially non-binding phrase, however, did not imply that the two parties had found a common understanding of the problem. Soviet perceptions of nature as either a means or an impediment to human aspirations – in the first instance to be exploited and in the second instance to be conquered[16] – were still alive in the Gorbachev era. As we will see in the following chapters, the establishment of the first Soviet environmental agency in 1988, the State Committee for Nature Protection (Goskompriroda), did not signify any substantial changes in the general Soviet approach to environmental protection. Even the eventual dissolution of the Soviet Union in 1991 – the state that fostered the notion of socialist conquest of nature – would do little to alter this.

The differences in approach existed not only on the level between Soviet and Norwegian bureaucrats. In Norwegian media and public opinion, the portrayal of environmental problems in the Soviet Union was becoming increasingly alarmist. This would, both in the short and long term, contribute to a stronger Norwegian commitment to environmental collaboration with the Soviet Union and later Russia. It would also, as we shall see, further demonstrate the existence of a deep divide between Soviet and Norwegian perceptions of how to best interact with the natural environment.

## The 'death clouds discourse'[17]

In the summer of 1988, prior to the first meeting of the joint Soviet–Norwegian Environmental Commission which was scheduled for late August, Pechenganikel – or the evident environmental degradation around Pechenganikel – became a hot topic in Norwegian news media. Covering the visit of Norwegian minister of the environment Sissel Rønbeck to Nikel in late June, Norwegian journalists relayed to their readers the first real glimpses into the previously hidden world behind the Iron Curtain. First impressions are said to be important, and in this case that is very true. The reports from Nikel would set the tone for Norwegian perceptions of life on the Kola Peninsula for years to come, and through numerous reproductions these perceptions would attain an almost hegemonic status in the Norwegian public discourse.[18]

A brief, and personal, visual description of Nikel may assist in the following discussion.[19] Anyone entering Nikel will be immediately struck by the physical presence of the combine factory buildings. The industrial complex situated on the northern edges of the village and its three towering smokestacks leave no doubt that this is a Soviet (now Russian) company town. The seemingly complete neglect of maintaining building façades over many years leaves an impression of an abandoned industrial site, countered only by the incessant roar from the smelting furnaces, the ever-rising columns from the smokestacks and the fact that workers are moving in and out of the factory gates. The non-industrial sections of Nikel give an equally bleak impression. Soviet apartment buildings are usually aesthetically displeasing to most Western eyes. They simply do not meet the culturally imposed expectations in most Western countries of what constitutes successful architecture. In addition, the almost brutal insertion of these urban multi-story dwellings in what was once pristine tundra moors makes them look misplaced and like scars on the landscape. The landscape itself, in the village centre and its immediate surroundings, is severely damaged. Black tree stumps protrude from heaped black mud no longer covered with the archetypical Arctic lichen and moss. A short car trip from Nikel, however, one can easily come across natural beauty seemingly untouched by industrial waste. The Kola Peninsula is, with the exception of high-pollution areas like Nikel, dominated by inaccessible outback areas with little human activity.

It was one of the heavily polluted sceneries that appeared before Norwegian journalists in late June 1988 when they were reporting from Rønbeck's encounter with the Pechenganikel combine management and the municipal administration in Pechenga.[20] As was to be expected, the resulting articles emphasized and amplified the shocking exoticness of a polluting industrial site in the wild tundra landscape. Under the heading 'Glasnost with a stench of sulfur', one Norwegian tabloid gave a vivid description:

> The sulfurous stench tears at your nostrils. Minister of the Environment Sissel Rønbeck coughs, the combine manager coughs, the deputy chairman in the Soviet environmental commission coughs. We are inside the main workshop in the Soviet border town Nikel: An unvarnished encounter with the cause of forest death in Finnmark. Sissel Rønbeck's visit to Nikel yesterday put the new Soviet openness to a difficult test. The Soviets brought her straight to the sulfuric acid production hall, in the heart of the nickel works – uncensored. . . . .
> [The emissions] have left scars. Square kilometers of forest around the mining town have simply vanished. Some bare stumps with straggling branches are the only remaining signs that the area was once green and luxuriant. Now the Nikel landscape is a sulfurous desert, and the invisible gas still streams in large quantities towards Norway. Pechenganikel emits 257,000 tons of sulfur dioxide annually, almost three times the total Norwegian discharge of sulfur dioxide. These are Soviet figures. The Norwegian ones are much higher.[21]

Despite the obviously theatrical tone of this article, the description of Nikel was fairly accurate. However, the apocalyptic-metaphorical language, such as the expression 'sulphurous desert' necessarily created one-dimensional images in the minds of the readers. It simply did not allow for the possibility of a fairly content resident population in Nikel going about their everyday lives: instead, it established Pechenga, and the Kola Peninsula as a whole, as a site of human suffering under a toxic spell. Furthermore, the 'invisible gas' that was allegedly threatening Norwegian territory fed fears of the dismal Soviet reality being extended westwards.

Similar reports were published in various Norwegian newspapers a little more than a month later, when the Norwegian Labor Party's youth organization, AUF, went on a trip to the Kola Peninsula, crossing it by train from north to south. AUF had teamed up with the Soviet youth organization KMO and visited Pechenganikel's smelters, among other places. According to one news report, the young Norwegian social democrats were 'shocked' to witness 'the enormous forest death on the Kola Peninsula'.[22] Another newspaper reported not only the astonishment of the Norwegian guests, but the equally alarmed reactions of Soviet youth delegates, who were apparently not northerners themselves. One KMO delegate stated: 'When we see Nikel's children breathing this gruesome air, and at the same time the dead trees around town, we have to ask ourselves why this is allowed to happen.' A Norwegian delegate followed up: 'What we have seen is nothing less than a crime committed against nature.' Dazed by the environmental degradation they had witnessed, the young politicians formulated a common declaration demanding an end to the misery.[23] No mention was made of the extensive stretches of untouched nature that the same young politicians must have passed through on their long train ride. This, presumably, did not fit with the general story line of the Kola Peninsula as an environmental disaster zone.[24]

When the Soviet–Norwegian Environmental Commission convened for the first time a few days later, its delegates were of course well informed of the hullabaloo in the press. The KMO and AUF declaration was noted, and the Pechenganikel emissions were very much the topic of the day. The Soviet delegation assured the meeting that considerable work was being done to curtail discharges from Pechenganikel, although this was arguably just a repetition of broken promises made since 1978 (see Chapter 2). While adamant about the realism in Soviet plans to limit emissions, the Soviet delegation agreed with the Norwegian delegation that border-area pollution would be discussed separately at future commission meetings.[25] However, when confronted with a group of inquisitive Norwegian journalists at the final press conference, Soviet delegation head Valentin Sokolovskii presented a somewhat mixed message. While assuring the journalists that pollution from Pechenganikel was being taken seriously by Soviet authorities, he offhandedly described it as 'a limited, local problem'.[26]

Sokolovskii had obviously miscalculated the effect of his statement. The rather laidback assessment of the environmental situation in Nikel would probably not have raised an eyebrow if presented to a Soviet audience. His words reflected the Soviet inclination to emphasize that, although admittedly blemished with some environmentally troubled settlements, Murmansk oblast was located on a peninsula of predominantly untouched wilderness. In the opinion of most people, local pollution was nothing to get 'all hysterical about'.[27] A Norwegian press corps preoccupied with overwhelming and disturbing images of a spreading sulphurous desert viewed the matter differently, and Sokolovskii's comment was brusquely dismissed in several editorials. His presumed lax attitude was seen as a sign that Soviet authorities were not approaching the problem with sufficient seriousness. Further, journalists commented, acid rain had already caused considerable vegetation damage to both Norwegian and Finnish territory. The problem was neither limited nor local, they argued.[28]

The Norwegian public certainly did not think the Pechenganikel emissions a trifling matter. That said, popular resistance to Soviet pollution was not a broadly national phenomenon but was mostly limited to the local communities directly exposed to emissions. People in Pechenga's Norwegian neighbouring municipality Sør-Varanger were deeply and increasingly concerned about the effects sulphur discharges might have on their immediate environs. In the spring of 1989, the chief municipal physician in Sør-Varanger presented a report that suggested there might be a correlation between the increased prevalence of cancer on the Norwegian side of the border and emissions from Pechenganikel.[29] With this, the problem became not only environmental but also a matter of public health. Naturally, this type of information stirred emotions on the Norwegian side. Sør-Varanger, a municipality in economic recession due to the fading fortunes of the cornerstone enterprise AS Sydvaranger, was already in a depressed state. Struck by high unemployment rates and claiming to be abandoned by central authorities in Oslo, the local population constituted fertile breeding grounds for activism directed against an outside enemy.[30] When in late 1989 and early 1990 several reports on the environmental state of the municipality and the dimensions of the Pechenganikel discharges were published, the stage was set for a highly emotional campaign directed against both the neighbouring Soviet polluters and the Norwegian authorities.

Four influential individuals with different political backgrounds and wide experience from organizational activities and specific environmental protests were instrumental in rallying the local community against what they labelled the 'death clouds' from Pechenganikel's smokestacks. Through a mixture of civil disobedience, including unlawful crossing of the Soviet–Norwegian border, skillful manipulation of the media and active use of their political contacts, these key figures were able to quickly recruit large numbers of supporters in the local community, elevate the Pechenganikel emissions high on the Norwegian political agenda and exert substantial pressure on authorities in Oslo. Albeit failing to achieve the same impact in Moscow or Pechenga, the action group did play a pivotal part in urging Norwegian politicians to design a scheme, discussed in the next section, for solving the sulphur dioxide problem.[31]

The emotional intensity of the Sør-Varanger activism was remarkable and reflected the urgency felt by its participants. At once both the main slogan and the name of the campaign, 'Stop the Death Clouds from the Soviet Union' clearly stated the objective as well as the perceived gravity of the situation. Sør-Varanger's battle against Soviet pollution was portrayed as a struggle for people's lives; if the discharges from Pechenganikel persisted, the consequences would be lethal. The campaign's spearhead Kåre Tannvik retrospectively commented on the strategy for arousing people's survival instincts: 'Appealing to people's feelings was an important aspect. Research reports will tell you a lot about pH values, but little about emotions. We brought to the surface the feelings that people had long repressed. That gave us a drive very few others had.'[32] For a limited period, this 'drive' was expertly utilized by the campaign leadership to feed news media with gloomy scenarios about the future of Sør-Varanger if emissions from Pechenganikel were not reduced substantially. The main impact was local, but the campaign managed to get coverage from a wide range of news agencies, both at home and abroad. Newspapers in Moscow, Murmansk and Arkhangelsk oblasts wrote about it. The media in Finland, Sweden, Denmark, Great Britain, Germany, and even the United States reported on the struggle against Soviet 'death clouds'.[33]

As we shall see, the campaign would in the short term achieve significant success and recognition, although achieving its ultimate target, a 90 per cent reduction in sulphur emissions from Pechenganikel, proved very difficult. Over time, the main impact of the 'Stop the Death Clouds from the Soviet Union' campaign and its reciprocal interaction with the media to launch and reproduce the narrative that Pechenganikel emissions would in a few decades turn Eastern Finnmark into an inhospitable desert[34] was to establish the 'death clouds discourse'.[35] For most of the 1990s, this discourse would strongly influence Norwegian public opinion about life on the Kola Peninsula and the threat it posed to Norwegian citizens and territory.

As illustrated by Valentin Sokolovskii's dismissal of Pechenganikel emissions as a 'limited, local problem', the discrepancy between Soviet and Norwegian perceptions of the matter at hand would only increase as the 'death clouds discourse' gained ground. In fact, irritation over what he perceived as hysterical news coverage of the environmental situation in the Pasvik valley led Sokolovskii to confront his Norwegian counterpart Oddmund Graham, asking him to 'monitor what is written in Norwegian press about this, and to correct misleading information'.[36] Various factors, apart from the obvious fact that the 'death clouds discourse' was brimming with overstated apocalyptic scenarios, prevented harmonization of Soviet and Norwegian perceptions of the Pechenganikel problem. Firstly, the Soviet perception of the Kola Peninsula as a fundamentally pristine wilderness tended to dampen the alarm. A certain degree of pollution was to be tolerated for the sake of productivity, even more so since the vast peninsula still offered many areas of untouched nature. Furthermore, the Pechenganikel emissions were, in a Soviet context, fairly moderate. In 1991 the largest nickel producer in the Soviet Union, the Norilsk Mining and Metallurgical Combine, emitted 2,400,000 tons of sulphur dioxide, whereas Pechenganikel's discharge was 195,000 tons – still substantial, but relatively modest by Norilsk standards.[37] Secondly, the fact that Pechenganikel was located on Soviet soil thwarted what little internal environmental opposition existed in the mono-industrial company towns of Nikel and Zapolyarnyi.[38] Not surprisingly, protests against pollution are lessened if the polluter happens to be your own employer, as shown by the reluctance of the residents of Sør-Varanger to fight their own employer/polluter AS Sydvaranger. Pechenga *raion* and Murmansk *oblast* found themselves, as parts of an unravelling USSR, in a progressively more precarious economic, social and political position by the late 1980s. As we have noted, the oblast's mono-industrial towns did not experience the same type of environmental activism as other, economically more diversified parts of the Soviet Union.[39] Environmental problems simply did not rank high in the problem hierarchy in the Soviet Northwest.

Nevertheless, Soviet bureaucrats were involved in an environmental commission where they had to respond to Norwegian pressure to find a way to curtail sulphur dioxide emissions in the Pasvik valley. As we shall see in the following paragraphs, Norway's involvement in the

The area surrounding the industrial site and town of Nikel is actively used for recreational purposes by local residents.
Photo: Ola Solvang

reconstruction of the Pechenganikel smelters was in fact a follow-up to negotiations that had already been going on for a while between Finnish and Soviet actors. Before long, also Sweden would join in to apply pressure on the Soviet Union.

## Nickels and dimes

Political pressure alone was insufficient. After all, Norwegian authorities had been trying to push for a solution to the sulphur emission problems ever since the late 1970s. This had not brought about the desired effect. Soviet authorities simply did not adhere to the 'polluter-pays principle' – an integral part of all international environmental agreements since the Stockholm Declaration in 1972 (including the LRTAP)[40] – but had allowed Pechenganikel's industrial processes to continue basically unchanged. On top of that, Norwegian interests had problems in convincing Soviet counterparts that Pechenganikel's industrial discharges constituted a problem at all. Obviously, a new impetus was needed to move the matter forward. Soon enough, the question of Western financial contributions to modernization of the nickel industry came up – first introduced by the Soviet side. In his outspoken book *Smokestack Diplomacy*, Robert G. Darst discusses the many implications a state's direct financial intervention in environmental problems outside its borders – what he calls 'environmental subsidization' – may have. Couched in direct language, his analysis points to the very real danger that international environmental efforts may be guided by completely different objectives than the protection of nature.

> In most cases in which polluters and the victims of pollution find themselves separated by international boundaries, the victims must rely on their own resources to persuade polluters to reduce transboundary environmental threats to an acceptable level. One possible response is direct financial incentive to reduce emissions or to improve industrial safety – in other words, a bribe.[41]

Although Darst here speaks about bribery in connection with the breach of the overriding 'polluter-pays principle', which would be handled on an aggregate non-individual level, there is reason to believe that financial motivations played a role also for Soviet individuals in the expanding East–West environmental cooperation. The industrial aspects and commercial opportunities were an often-overlooked motivation for Western partners, in this case Finland and Norway, as well.

One of the institutions reflecting Finland's close ties to the Soviet Union after the Second World War was, since its establishment in 1967 until fall 1991 (when it was abolished – see later in this chapter), the Finnish–Soviet Intergovernmental Commission for Economic Cooperation (hereafter the Finnish–Soviet Economic Commission).[42] Chaired by influential political actors from the two countries, this body developed into a powerful instrument in the comprehensive Finnish–Soviet clearing trade in the post-war period.[43] When in the

1980s the nickel industry on the Kola Peninsula required upgrades, Finnish industry made its services available for Soviet customers and the Economic Commission became a framework for potential modernization projects in the Soviet non-ferrous metallurgical sector. These projects were initially commercially motivated – the Soviet side aimed at maintaining its status among the world's foremost nickel producers, while Finnish industrial actors saw an opportunity to land lucrative long-term contracts.[44] After Gorbachev gave his speech in Murmansk in October 1987, high-level Soviet and Finnish politicians on several occasions discussed the modernization of the nickel works on the Kola Peninsula for reasons that included environmental concerns. In November 1988, Finnish prime minister Harri Holkeri and his Soviet colleague Nikolai Ryzhkov established a Finnish–Soviet working group to develop technological solutions for the reconstruction of both Pechenganikel and Severonikel in Monchegorsk with the aim of reducing emissions. Gorbachev discussed the matter further with Finnish president Mauno Koivisto in October 1989.[45] The Finnish metallurgical company Outokumpu OY had by then already entered negotiations with the Soviet authorities for the complete renovation of the smelters in Nikel and Monchegorsk (Severonikel) as well as the modernization of installations in Zapolyarnyi.[46]

Soviet actors did not limit themselves to Finnish counterparts, however. A prospective modernization was also discussed within the framework of the Soviet–Norwegian Environmental Commission. At its first session in August 1988, the commission called for a meeting between Norwegian and Soviet experts on discharge purification.[47] When Norwegian representatives arrived in Moscow in February 1989, Head Engineer Igor Borodin in the Soviet ministry for non-ferrous metallurgy's (Mintsvetmet) ecological branch *Tsvetmetekologiya* invited them to develop technological solutions to the emission problems on the Kola Peninsula. If this was done, guaranteed Borodin, purification installations would be built according to Norwegian specifications.[48] Borodin's bold guarantee made an instant impact on the Norwegian representatives. At the Environmental Commission's second session in

April 1989 the parties agreed that a Norwegian expert delegation should be sent to Pechenga to 'assess the possibilities of applying Norwegian technology at these installations'.[49]

Very soon, preparations for expert exchanges were underway. Already in March 1989, the Norwegian firm Elkem Technology visited Pechenganikel.[50] Also, Norsk Hydro and Falconbridge (Canadian-owned company located in Kristiansand, South Norway) were invited and expressed interest in participating in a visit to Nikel as well as in receiving Soviet experts at their own installations. While the Soviet experts were expected to inspect advanced Norwegian environmental technology, the Norwegian visit to Nikel was part of a plan for developing technological solutions to the Pechenganikel emission problems.[51] Amidst all this, Norwegian and Soviet environmental bureaucrats kept in touch. In mid-August 1989, the heads of the delegations to the Soviet–Norwegian Environmental Commission had consultative talks in Moscow. Especially interesting was the exchange between Valentin Sokolovskii and his Norwegian colleague Oddmund Graham during the 'informal' portion of the meeting. First, Sokolovskii 'confidentially' suggested the establishment of a joint Soviet–Norwegian technology development company to provide emission reductions at Pechenganikel. Thereafter, he introduced another and, as time would show, a more realistic possibility:

> While pointing to what has been written in Norwegian press about the need for aid to solve environmental problems in Eastern Europe, especially in Poland, Sokolovskii suggested that the Norwegian side might want to consider financial assistance to emission reduction at the nickel plants by the Soviet–Norwegian border.[52]

Graham replied hesitantly to both suggestions. He left the idea of a joint company to the experts who were to meet in both Nikel and Norway shortly thereafter. In response to possible Norwegian assistance to Pechenganikel, he pointed out that the Soviet Union was in a better position than Poland to deal with its own environmental problems, but otherwise left the question unanswered. Be that as it may, Sokolovskii had let the cat out of the bag: if emissions from Pechenganikel were going to be curtailed, Norway would probably have to contribute both financially and technologically.

The opening up to Norwegian interests, and the intimations about Norwegian subsidization in particular, had several implications. Firstly, more emphasis was put on the installations in Pechenga than on the Severonikel plant in Monchegorsk. For a while, the modernization plans still comprised the whole of the nickel industry on the Kola Peninsula, of which Severonikel was part. However, Monchegorsk, located at a safe distance from the Norwegian border, was no longer defined as a prime target, at least not by the Norwegians. Secondly, and connected to this, the environmental profile of modernization was strengthened. Although industrial aspects were still important, the primary Norwegian goal was, at least in rhetoric, to achieve radical reductions in sulphur dioxide emissions. In the public sphere, the modernization of Pechenganikel was mainly portrayed as a project for reducing pollution. Finally, the Western neighbours were given a first glimpse into what was to become a Soviet and later Russian approach to the modernization project. By inviting a new actor – the Norwegian government – onto the playing field, the Soviet side was able to raise interest in the problem at hand. Thus, the pollution from Pechenga became more of a bargaining chip than an international problem that had to be solved. While the smokestacks were still spewing out sulphur dioxide, the polluter had leverage over potential financial contributors. Pechenganikel emissions became a resource for increasingly cash-strapped Soviet and Russian actors.

This development should be seen in the context of internal processes in the Soviet Union at the time. Sokolovskii's idea of a Soviet–Norwegian technology company was most likely inspired by the law on 'joint ventures', adopted in January 1987. Briefly put, this law allowed for Soviet enterprises and cooperatives to enter into partnerships with foreign companies seeking to do business in the Soviet Union. Attractive as this opportunity might have seemed to foreign companies, restrictive legal measures ensured that profits from and control over such 'joint ventures' stayed mostly in Soviet hands. This and the fact that the Soviet partners were principally interested in access to advanced technology, investments and hard currency, rather than mutual value creation, ensured that many Western companies chose to refrain from entering into such agreements.[53]

Another legal change was more important to Pechenganikel. In June 1987, the Law on State Enterprises was passed as part of Gorbachev's attempt to raise industrial efficiency. The law provided, at least nominally, for greater autonomy in state enterprises, by relieving them of the burden of being subjected to strict vertical control of the industrial ministries in Moscow. The hope was that greater flexibility would enable managers and employees, who were also given direct influence on executive decisions, to better exploit the common resources of the enterprise. Although the intention of the law was systematically subverted by the ministries,[54] Mintsvetmet among them, the law itself was a strong indication that established power structures were being challenged. In times of change, those with ambitions will seek to position themselves strategically in order to gain ground in the struggle for future control. This was true also in the case of the Soviet nickel industry.

The consolidation process that culminated in the creation of Norilsk Nikel, a state-owned concern, in November 1989, of which Pechenganikel was part, implied only a pause in the unsettling reshuffling of power.[55] Actual stabilization in Soviet and post-Soviet industry would, due to economic instability, not be achieved for a long time. That being said, it was in this first period of upheaval, when command lines and responsibilities on the Soviet side were unclear, that the modernization of Pechenganikel was first introduced as a possible multinational effort. In attempting to position themselves as negotiation counterparts to Western actors, several Soviet (and later Russian) individuals would make assertions and pledges that were hardly within their power to make. It is tempting to see the abovementioned guarantee from Borodin in Tsvetmetekologiya, and Sokolovskii's intimations about a joint Soviet–Norwegian venture subsidizing the Pechenganikel modernization as early examples of this.

What, then, created an environment conducive to such unauthorized personal initiatives? The industrial chaos was mirrored in society at large: the attempted restructuring of the Soviet production system was accompanied by a dramatic downturn in Soviet economy. By 1989, the many internal weaknesses of the planned economy were becoming alarmingly evident, and the Soviet self-perception of a mighty socialist state with zero unemployment and abundant welfare services was crumbling rapidly. This was not only felt on the macro-economic level, but palpably affected individuals and families in their everyday lives. More and more Soviet citizens were stricken by poverty – their numbers varying according to the statistical methods applied. There was no denying that the socialist experiment was in dire straits. In October 1989, Deputy Prime Minister Leonid Abalkin claimed that over 20 per cent of the Soviet population could be described as poor – a figure he considered a disgrace to any civilized, socialist country.[56] Although not necessarily among the most unfortunate, few prospective Soviet participants in East–West cooperation were able to evade the crisis on a personal level. Arguably, then, already at this point the lucrative aspects of collaborating with Western countries became a motivating factor for Soviet individuals who were in a position to do so. In the Soviet–Norwegian context, a rather handsome daily allowance (in exchangeable hard Norwegian currency) was paid to Soviet delegates while staying in Norway, in addition to food and lodging expenses.[57] Later on, as we shall see, larger sums would be at stake.

To sum up, Soviet–Norwegian talks regarding the modernization of Pechenganikel were initiated in a period of massive inner turmoil in the Soviet Union. The nickel industry, like most other Soviet industries, was in a process of fundamental change that brought the command structure in disarray. This opened up for individual and sometimes unsubstantiated initiatives from actors within the Soviet nickel complex who were trying to gain ground in the ongoing struggle for control in the industry. By appearing sympathetic to Western ideas, one could hope to achieve an advanced position in future negotiations with affluent neighbours who were, irrespective of environmental or industrial motivation, able to fund comprehensive projects. On the more personal level, merely participating in East–West cooperation gave increasingly impoverished Soviet bureaucrats access to hard currency and other perks that were otherwise unattainable. Ultimately, in the case of Pechenganikel, it was the sulphur emissions that gave Soviet and later Russian actors these possibilities. In that sense, the emissions themselves became a resource rather than a problem. It is also against this background that the story of modernization in Pechenganikel must be understood.

## Nickels and dimes II

Subsidiary or non-environmental motivations were not restricted to the Soviet side of the border, however. The commercial aspects associated with modernization of the nickel industry on the Kola Peninsula were extremely important also to Western actors and would play a pivotal role in shaping their engagement in the matter. This was, as mentioned, obvious in the Finnish case. Outokumpu's interest in the modernization project had an openly industrial and primarily commercial character, developed as it was within the framework of the Soviet–Finnish Economic Commission.[58] In the Norwegian case, by contrast, the commercial aspects were decidedly under-communicated. When Prime Minister Jan P. Syse in September 1990 declared his government's decision to contribute 300 million NOK to the modernization efforts, he depicted it as a measure to 'save the natural environment in Finnmark'. Employing elements from the 'death clouds discourse', he went on:

> The desert is spreading on the Kola Peninsula. . . . The development is clear: Sulfur and heavy metals are exerting an unbearable strain on Finnmark's nature. We cannot accept this. We will not accept this. This development must be reversed. We will participate in the reversal. This is what the government has ordered. The path may be long and difficult. No one has walked it before us. We will protect the environment in

Norway by approaching the source of pollution, beyond our own borders. The special thing about this case is that damages in Finnmark cannot be remedied through measures in Finnmark. Measures must be taken on the Kola Peninsula.[59]

Although Syse mentioned that the modernization also could serve as a 'springboard for expanded economical, commercial and cultural contacts', that came as more of an afterthought. His audience that September evening – the speech was given at a gala concert in Oslo Concert Hall organized by 'Stop the Death Clouds from the Soviet Union' – dictated an emphasis on the environmental aspects. He was probably closer to the essence of the matter when he stated that 'this is practical Nordic cooperation. Together we will make use of Nordic funding mechanisms. Together our enterprises will work to deliver the best possible technology'. It was mainly these two issues – who would fund the modernization project and with how much, and who would provide the technology – that preoccupied Norwegian and Finnish actors at the time.

In the months prior to Prime Minister Syse's grand gesture in Oslo Concert Hall, the Norwegian government bureaucracy started the work of developing a modernization scheme. Two issues stand out in the comprehensive correspondence concerning the matter: How to fund modernization and how to maximize the benefits for Norwegian contractors and other businesses that might play a role in this considerable industrial endeavour. The environmental aspects were of course important – it was after all the environmental degradation around Pechenganikel that had raised the Norwegian interest in the first place – but they seem to have been quickly overtaken by technical considerations and commercial aspirations. The original focal point, solving a major environmental problem in the far north, became blurred by other aims that would complicate the road towards a final solution to the sulphur emission problem. As we shall see, there were plenty of economic considerations – which had nothing to do with the northern environment as such – that would stand in the way of expedient modernization of the Kola Peninsula nickel industry.

First of all, getting rid of the sulphurous emissions was not a simple matter of putting 300 million NOK (about USD 46 million) on the table.[60] Outokumpu's project plan, which involved full modernization of the Kola nickel smelters in both Nikel and Monchegorsk, envisaged costs between USD 600 and 750 million,[61] which meant that the Norwegian contribution would amount to only a minor portion of the total value of the project. Although Finland – provided that Outokumpu's project was chosen – was prepared to contribute up to twice as much direct financial support as Norway,[62] the modernization would still lack about five-sixths of the required funding. There was little help to be had east of the border. As Sokolovskii had indicated, the Soviet Union would not be able or willing to make this type of investment in its industry – an investment that would have little impact on the industrial efficiency or the profitability of the nickel works and was deemed necessary only by Western neighbours who were purportedly motivated by environmental considerations.[63] The triumphant mood among the environmental activists in Oslo Concert Hall after Syse's announcement must therefore be described as premature.

Inasmuch as the combined Finnish and Norwegian direct contributions, ranging anywhere from USD 90 million to almost USD 140 million,[64] would in any case cover only a fraction of the total costs of Outokumpu's project, other sources of funding would have to be identified. Throughout 1990, Finnish and Norwegian bureaucrats worked together to design an elaborate funding scheme. Both governments were bound by agreements reached in the OECD that barred member states from giving direct gifts, or what were called 'interest rate subsidies', to the crisis-ridden Eastern European countries.[65] In effect, this meant that the Finnish and Norwegian contributions would have to be channelled through a third party. The solution most frequently discussed was to set up a fund to be managed by the NIB, which had been established in 1975 by the Nordic countries (Finland, Denmark, Sweden, Iceland and Norway).[66] Finland and Norway also involved the Swedish government, in the hope of attracting more funding. The Swedish government hesitantly expressed willingness to contribute but was non-committal with regard to the size of the sum.[67]

Again, direct support from the Nordic partners would take the project only a small part of the way towards full funding. Most of the required money would have to be raised as loans – loans that were expected to be paid back in full, in this case by Norilsk Nikel, a seemingly broke debtor. Furthermore, the increasingly turbulent economy in the Soviet Union was making Soviet loan guarantees progressively more untenable. Not only was the solvency of the state considered questionable, the very future of the USSR was in doubt. Consequently, Western financial experts recommended seeking guarantees not only from the Soviet Union, but also from the RSFSR.[68] The Soviet market was considered extremely risky both by Western companies and Western guarantee institutions. The Norwegian institution for export credits (GIEK) was apprehensive,[69] as was the Norwegian export funding agency Eksportfinans, about the risks associated with entering the Soviet market.[70]

This apprehension was shared by companies aiming to get involved in the modernization project. The Norwegian firm Elkem Technology, whose interest in Pechenganikel will be discussed in this chapter, insisted that Norwegian authorities cover the lion's share of their development costs, as they assessed 'the political risks involved in accomplishing this project to be unusually great'.[71] In addition came the justified worries about Soviet solvency. Motivated by prospective profit, the Western companies were aware of the lack of hard currency in Soviet enterprises and knew that Norilsk Nikel was in no position to pay for the modernization in cash. Part of the funding, therefore, was foreseen to be arranged through an exchange of refined nickel against Western technology. The major Finnish and Norwegian industrial actors were amenable to such an arrangement – but it presupposed that Norilsk Nikel would be granted an export licence from the Soviet state planning agency Gosplan for a product that was traditionally reserved for domestic consumption.[72]

As we see, Norwegian prime minister Syse was correct when he stated in September 1990 that 'the path [towards modernization] may be long and difficult'. Technical hindrances to this multilateral venture abounded, as did the risk factors, and the complexity of the funding scheme could in itself be prohibitive to any contractor aiming to get involved. Nonetheless, Finnish and Norwegian companies were eager to participate, viewing the modernization venture as an early opportunity to gain a foothold in the Soviet industrial market. In Norway, this enthusiasm was shared by the government. The key factor was choice of technology. Even environmental bureaucrats emphasized the need to ensure that proven Western technology – and most importantly, *Norwegian* technology – would be installed in the nickel smelters. Again, the argument was primarily environmental: Norwegian and Finnish solutions were held to be superior to Soviet technology in terms of emission reductions, so both Finland and Norway insisted that their technology be used in the modernization. However, purely commercial motivations also played an important part in forming this stance.

As mentioned earlier in the chapter, it was initially the Finnish company Outokumpu alone that was involved in the modernization plans being developed under the auspices of the Finnish–Soviet Economic Commission. A supplier of stainless steel, Outokumpu had an interest in gaining access to nickel, the scarcity of which was a bottleneck in their production.[73] However, projected costs related to Outokumpu's modernization, which was developed as a wholesale rebuilding of all industrial facilities in both Nikel and Monchegorsk, grew to a prohibitive level as planning moved ahead. The price increase coincided with an emerging realization in the Soviet Union that the country was no longer able to carry such economic burdens. Thus, Finnish–Soviet negotiations lagged, and it became clear that the project would be unfeasible as a purely commercial venture to be paid for by the Soviet state. Still intent on seeing the modernization plans come to fruition, Finland in October 1989 suggested another solution: if Finnish technology (i.e. the Outokumpu project) was chosen, Finland would provide intermediate funding for the modernization, to be paid back by the Soviet state on terms to be agreed. However, that did not assuage Soviet apprehensions, and it became increasingly likely that Norilsk Nikel would end up choosing domestic technology for the modernization of the Kola nickel industry.[74]

Norwegian authorities were privy to this information through the Soviet–Norwegian Environmental Commission. Worried that Norilsk Nikel would choose domestic Soviet technology before having assessed Norwegian solutions, the Ministry of the Environment encouraged the Norwegian company Elkem Technology to come up with an alternative plan for modernization.[75] Elkem Technology had, as mentioned, been among the Norwegian companies to visit Pechenganikel already in March 1989, aiming to provide environmental technology there, and in late 1989 had established a project group for the purpose. This group made several further visits to Pechenga and established contacts with the combine management. Elkem Technology quickly concluded that it was well equipped to solve the emission problems and that its solutions were in every way better than both Outokumpu's modernization plans and the domestic Soviet solution based on the so-called Vanyukov furnaces.[76] Not only could Elkem Technology, in its own view, perform equally well or even better in terms of emission reductions, it would also be able to do this at a substantially lower price. Elkem Technology dismissed the Soviet plans as insufficiently beneficial to the environment – but, even more importantly, the Soviet Vanyukov technology was according to them underdeveloped and far too costly for the impoverished Soviet state concern to fund. Nevertheless, Elkem Technology signalled flexibility in stating that it was prepared to act as subcontractor to either of the two other alternatives, should the Norwegian solution not be favoured. In referring to the 'massive political and economic support' Outokumpu received from the Finnish government, however, Elkem Technology insisted that Norwegian authorities level the playing field by providing them with the same assistance.[77] The support was forthcoming, and Elkem Technology on

20 June 1990, received from the Norwegian government a handsome project development endowment equivalent to USD 400,000.[78]

The situation halfway through 1990, then, was at a standstill: Outokumpu and Elkem Technology, two companies with strong ambitions in what they saw as a profitable project and an entry to a future 'huge market' for 'environmentally friendly' technology,[79] were courting a customer that was in no position to pay for their services. Norilsk Nikel, on the other hand, was developing its own technology. However, these Soviet plans seemed equally unrealistic due to high costs.

Two changes would break the standstill. Firstly, Outokumpu and Elkem Technology decided to join forces – although this union was hardly a joyful one. In an agreement signed 22 August 1990, the companies declared their wish to 'explore ways to cooperate in the reconstruction' of 'the Pechenga Nikel [*sic*] smelting plant (and possibly also Severonikel primary nickel smelter capacity)'. Outokumpu was given a leading role in the collaboration, whereas the exact status of Elkem Technology would be decided at a later stage.[80]

Seen from Outokumpu's side, this arrangement was probably more one of political necessity than of industrial and commercial convenience. As Norwegian authorities had become increasingly clear about their intention to contribute financially to the modernization, Norwegian contractors followed suit and claimed, with governmental backing, what they saw as their rightful share of the project. In Outokumpu, there was a certain degree of irritation with what was seen as Elkem Technology's encroachment on their project. Outokumpu CEO Pertti Voutilainen, despite the arrangement with Elkem Technology, later stated that as much work as possible should be done by Outokumpu, to ensure unity and control of the project. That there was a basically commercial motivation behind this statement was obvious to the newspaper that conducted the interview with Voutilainen. The article's heading was 'Filatov [chairman of Norilsk Nikel] saves Kola Peninsula nature – Voutilainen does business: Nickel project jackpot for nature – and for Outokumpu.'[81]

The second factor that would move the modernization process further was a Norwegian government initiative. As the Outokumpu–Elkem deal was preparing the ground for more solid political unity among the Northern governments, Prime Minister Syse called a meeting. In connection with a Baltic Sea conference in Ronneby, Sweden, in early September 1990, Syse suggested that the prime ministers of Finland, the Soviet Union, Sweden and Norway meet to discuss possible emissions reductions from the nickel plants on the Kola Peninsula.[82] This meeting was held only days before Syse announced in Oslo Concert Hall his plans to fund the modernization. His message in Ronneby, supported by Finland and Sweden and communicated to Soviet deputy prime minister Vitalii Doguzhiev, who attended the meeting in place of Nikolai Ryzhkov, was unambiguous:

> It is our clear opinion that only a solution based on Western technology will bring the emissions down quickly enough (95% reductions within the mid-1990s). In the current economic situation, it seems obvious that such a technological solution will only be possible if the Soviet Union is offered a broad economic and technological cooperation from the outside world.[83]

With this, the Soviet side was confronted with a fait accompli: only Outokumpu and Elkem Technology's solution was deemed acceptable and only by choosing this would the Soviet state receive financial support for the modernization project. To Deputy Prime Minister Doguzhiev, the meeting with his Nordic peers must have been somewhat disturbing, although not atypical of the era: the previously mighty but now weakened Soviet Union was offered aid packages from small northern neighbours on predetermined terms. Doguzhiev readily admitted that the Soviet Union for many years had neglected environmental problems and that the time had come to pay the price, but he also pointed out that the company towns on the Kola Peninsula were not the end of the problem. According to Doguzhiev, Nikel and Monchegorsk were two among many – over one hundred – Soviet cities that were defined as areas struck by 'ecological catastrophe'. The deputy prime minister had to admit that the modernization project was not within the Soviet Union's reach in the economic situation current at the time.[84]

For the Norwegian government, on the other hand, funding was within reach. Moreover, it was highly desirable that the modernization project be realized – not only for ecological reasons. As mentioned in the preceding paragraphs, Pechenga's Norwegian neighbour municipality Sør-Varanger was in the early 1990s in a deep recession due to the demise of the cornerstone enterprise AS Sydvaranger. The Norwegian authorities, and in this case the Ministry of the Environment, saw in the modernization plans an opportunity to revitalize the town of Kirkenes, which was the municipal centre of Sør-Varanger:

> If a reconstruction of the nickel plants near the Norwegian–Soviet border, in the form of a collaborative effort that is now being discussed, goes ahead, Kirkenes will be a potentially important supply hub for the project. It is important to emphasize the economic effects and creation of new employment both during the construction period and in the long run.[85]

Evidently, the purely economic dimensions, in this instance related to the revitalization of Sør-Varanger, of what was ostensibly an environmental project were important not only to the many Norwegian contractors that hoped to play a role in the industrial adventure, but also to the Norwegian environmental bureaucracy and politicians. The potential role of Kirkenes as a loading and accommodation area was of particular interest to local entrepreneurs, who wanted to make sure that Norwegian authorities applied pressure on Outokumpu and the Finnish authorities to secure contracts for them.[86] True, some quarters within the Norwegian bureaucracy were wary of possible negative effects of mixing commercial considerations with the overriding environmental character of the project and warned against letting Norwegian private enterprises 'abuse the public will to better the environment'.[87] However, these considerations soon proved to be of lesser significance. What happened next was that a dramatic chain of events, totally unrelated to sulphur dioxide emissions, came to threaten the clean-up on the Kola Peninsula. Unforeseen developments in the Soviet Union would fundamentally change the game.

## USSR vs. RSFSR

In March 1990, Gorbachev's foremost rival from within the ranks of the Soviet Communist Party, Boris Yeltsin, was elected chairman of the RSFSR Supreme Soviet. This made Yeltsin the most prominent politician in the most prominent Soviet republic and in effect gave him a platform from which he could challenge the general secretary. Not long after, on 12 June 1990, the Russian Federation proclaimed its sovereignty. An expression of the republic's desire to be freed from its oppressive USSR superstructure, the declaration provided for supremacy for Russian laws over Soviet laws and underscored Russia's right to leave the Union if its representatives so wished. Among the Soviet republics, RSFSR had historically enjoyed the least autonomy. Seen as a potential breeding ground for Russian chauvinism, the republican agencies of the RSFSR were fewer and less powerful than those in other republics. On 12 June 1991, exactly one year after Russian sovereignty had been declared, Boris Yeltsin was elected the first president of the RSFSR. Paradoxically, his ensuing rise to the pinnacle of Russian politics was aided by his most fervent adversaries, the conformist bloc in the Soviet Communist Party, who in August that year attempted to seize power by force. Skillfully outmanoeuvring the conservative coup-makers and dictating post-coup proceedings, Yeltsin asserted his position as Russia's leading political figure. Ultimately, Yeltsin dealt the Union of Soviet Socialist Republics the final blow on 8 December 1991, when he met with Ukrainian president Leonid Kravchuk and Belarussian leader Stanislav Shushkevich to form the Commonwealth of Independent States, thereby rendering the USSR superfluous.[88] The Soviet Union was officially dissolved on 25 December 1991.

Of course, this straightforward storyline, composed with the benefit of hindsight, does not reflect the many intricacies of the final Soviet struggle. Power shifted back and forth, and very few could predict the final result. A dominant conflict dimension was the antagonism between the republics and the union level. Like most of the other Soviet republics, RSFSR was striving for greater autonomy – a struggle that found expression in many contexts. Also, in matters concerning the nickel industry on the Kola Peninsula, which was located on the territory of the RSFSR, the antagonism between union and republic was felt. This power struggle, which for a long time blurred responsibilities, became a major impediment to progress in the modernization project. To Finnish and Norwegian actors, there emerged a central but often unanswerable question: with whom should they negotiate?

Already in the early autumn of 1990, the role of the RSFSR in the modernization project was discussed in meetings between Norwegian and Soviet actors and RSFSR representatives. After the proclamation of sovereignty in June that year, the Russian government became increasingly active and assertive in policy areas pertaining to the RSFSR and negotiated with the Soviet Council of Ministers for control over industrial and financial resources.[89] In this situation, the NME was eager to establish to what extent Russian republican authorities would be involved in further development of the modernization plans.[90] When, at the end of September 1990, the Norwegian ambassador to the USSR broached the subject in separate conversations with RSFSR deputy prime minister Igor Gavrilov and Soviet diplomat Yuri Fokin, reactions predictably varied. Whereas Gavrilov emphasized that the RSFSR should have full access to all proceedings in the project and proposed that Western aid be channelled through his government, Fokin downplayed the potential role of the RSFSR, stating that the republican government was very much in an early phase and still fully dependent on the Soviet apparatus for information.[91] For the time being, Fokin's assessment appeared the most reliable. Despite their waning power, Soviet authorities were still the main counterpart of

Finland and Norway in the modernization project, whereas the Russian government was merely kept abreast of developments.[92]

Albeit still in a leading role, the Soviet authorities seemed incapable of making the necessary decisions to move the project forward. Outokumpu and Elkem had presented a project plan to Norilsk Nikel in mid-November 1990, with several adjustments later that same month, but the all-important Soviet reply was not forthcoming.[93] The few signals that reached the Norwegian and Finnish actors indicated a downsizing of the project. Given the substantial price tag, modernizing both Pechenganikel and Severonikel in Monchegorsk did seem unrealistic. Pechenganikel, as the heaviest polluter and situated closest to an international border, was considered the best option.[94] This downsizing received reserved support from the regional authorities in Murmansk,[95] who were also becoming increasingly active in the spring of 1991. In March, leaders of the Russian Northwest met with the Norwegian ambassador, and a Norwegian embassy employee toured Murmansk oblast for talks with central political actors there. The general message received by the Norwegian embassy was one of defiance. The regional actors had little or no trust in the Soviet authorities and were also disappointed when it came to Yeltsin's general political accomplishments. Claiming a role for themselves, the regions sought direct collaboration with the Scandinavian countries and saw the modernization project as a central component of this.[96]

A project the size of the nickel industry modernization, however, was always a matter to be dealt with on the central level. But in this case, the 'central level' became increasingly elusive as union-level authorities dwindled and RSFSR authorities were still in no position to replace the waning Soviet power structures. Throughout most of 1991, it was very difficult for Norwegian actors to get in touch with any authoritative Soviet political figure. Prime Minister Gro Harlem Brundtland did consult with Gorbachev during the latter's visit to Oslo in June 1991 for a belated presentation of the Nobel Peace Prize he had been awarded the previous year.[97] As it turned out, however, Gorbachev and Brundtland's ambition of having the modernization completed by

1995 was unrealistic. Gorbachev's commitment to this completion date was to have scant significance, as his days at the helm of the unravelling union were numbered.

Also, in the main Soviet counterpart to Norway's Ministry of the Environment, the situation was as unclear as before. In July 1991, rumour had it that the State Committee for Nature Protection (Goskompriroda), established in 1988, had now been dissolved. Goskompriroda's replacement, the Soviet Ministry of Nature Protection (Minpriroda), was still not active. The RSFSR State Committee on Ecology (Goskomekologiya) was also going through a restructuring process.[98] The ensuing confusion between these two disorganized agencies about areas of competency made progress in the modernization project even more difficult. Everything was up in the air. In August 1991, a group of Soviet delegates managed to meet with their Norwegian colleagues,[99] but were not able to clarify the muddled future. In September, the NME was keen to meet with environmental authorities on both the union and republican level to clarify the division of labour between them.[100]

While the Soviet political scene was disintegrating, the modernization project seemed to live a life of its own in industrial circles in Finland and Norway. Negotiations between Outokumpu and Elkem Technology continued, although the partnership was fragile. Outokumpu was intent on keeping as many subcontracts as possible in Finnish hands and was reluctant to include Elkem Technology as a full partner.[101] Nevertheless, the two companies jointly presented a new and adjusted offer (involving modernization only in Pechenga) to Norilsk Nikel in late October 1991, with an estimated cost of USD 600 million.[102] Many smaller enterprises were also trying to position themselves for future subcontracts. The Norwegian authorities frequently received letters from industrial actors requesting political support for bids to take part in the modernization project.[103] Comprehensive plans were outlined, and many work-hours put in, to prepare for an industrial venture that in reality was extremely tentative, and that would take place in a state that was nearing disintegration.

A further step towards the complete collapse of the Soviet Union was taken in October 1991. Following haphazard reorganizations in the Soviet ministerial structure, several Soviet ministries were simply dissolved. Arriving in Moscow to discuss further plans for the modernization project, Finnish representatives to the Finnish–Soviet Economic Commission were informed that their counterpart, the Soviet Ministry for Foreign Trade, no longer existed, and that matters from now on would be handled by RSFSR agencies. However, no one was able to explain just who the Finnish delegates should approach in the Russian republican apparatus. In an attempt to clarify this, the Finns had discussions with the RSFSR's Goskomekologiya, RSFSR's Supreme Soviet and the general secretary for the now disbanded Finnish–Soviet Economic Commission. All expressed interest, but none could claim responsibility.[104] Later that same month, the NME was notified that their Soviet counterpart would cease to exist from 1 November 1991.[105] Only defence, transport, nuclear energy and state security agencies would be upheld at the union level. In other words, negotiations on modernization of the Kola nickel industry would have to start all over again – this time with the RSFSR.[106]

Spurred on by an impatient public opinion,[107] the NME was eager to get the modernization project back on track as soon as possible. Gro Harlem Brundtland seized the opportunity to meet with RSFSR's deputy prime minister Igor Gavrilov when the latter visited Oslo in late October 1991 for a business convention. As an illustration of how wide open the realm of possibilities was at this time, Gavrilov now suggested that Norwegian and Finnish industrial interests should take over the reins at Pechenganikel, and thereby introduce environmental improvements as owners. The Norwegian prime minister was in no position to answer for Norwegian entrepreneurs and pressed instead for a swift Russian decision on technology – that is, a response to the offer from Outokumpu and Elkem Technology. Igor Gavrilov agreed to the need for this and identified the RSFSR's Goskomekologiya as the responsible authority.[108] State Secretary Jens Stoltenberg was able to meet with Russian environmental authorities remarkably soon. On 7

November 1991, only one week after the formal dissolution of Soviet environmental agencies, representatives of the RSFSR's recently formed Ministry of Ecology and Natural Resources (Minekologiya – formerly Goskomekologiya) arrived in the North Norwegian town of Tromsø for discussions.

Liya Shelest, deputy minister of minekologiya, presented a mixed message. She immediately put to rest all worries about the future environmental cooperation between Russia and Norway, and was adamant that all working groups under the Soviet–Norwegian Environmental Commission would continue, now as Russian–Norwegian efforts.[109] She even expressed great interest in examining possible unsecured nuclear waste on the Kola Peninsula and the adjoining marine areas – the very existence of which had been categorically denied by the Soviet Commission chairman Valentin Sokolovskii.[110] When it came to the modernization project, however, Shelest and her colleagues were far less forthcoming. For one thing, the complexity and costs related to modernization meant that a swift decision was unrealistic, and the Russians demanded another three months to evaluate the industrial proposal.[111] More surprising, however, was Shelest's take on the viability of the project:

> We must assess the nickel plants on the Kola Peninsula in relation to the future supply of raw material and the metallurgical industry in the country as a whole. The industry along the Norwegian border might be shut down twenty years from now, and in that case large investments here will be nonsensical.[112]

Shelest's reference to 'the future supply of raw material' was meant to suggest that Pechenganikel's ore base was waning, or at best uncertain. The need to see the future of Pechenganikel in light of 'the metallurgical industry in the country as a whole' hinted at Pechenganikel's lesser role as a smelting facility within Norilsk Nikel. In sum, Shelest clearly indicated that the industrial future of Pechenganikel was uncertain and that massive reconstruction schemes at its installations might be a waste of time and money. Thus, she for the first time expressed fundamental doubts about the usefulness of the modernization project. Was it in fact necessary to expensively upgrade an industry which had a limited lifespan? Whether it was prophetic clarity or Soviet anti-hysteria discourse that inspired Shelest's comment, it was clear that Russia would be no easier to negotiate with than the Soviet Union had been.

*   *   *

The first years of Soviet–Norwegian environmental cooperation were eventful. Many collaborative areas were developed, and the modernization of the nickel industry on the Kola Peninsula was the most spectacular scheme among them. However, the two sides' understanding of the problem did not quite match. Environmental protection was traditionally not a priority in the Soviet Union, whereas the Norwegian environmental bureaucracy was confident in its approach to industrial enterprises, applying legal measures to halt unlawful pollution. Popular depictions of the problem differed greatly in the two countries. While the Norwegian public was told, by activists and news media, about a potentially disastrous influx of 'death clouds' over Norwegian territory, Soviet perceptions of nickel industry pollution were tempered by a wider context. First of all, the polluting factories were major employers and providers of important nickel, not merely a sulphur-emitting nuisance. Secondly, Pechenganikel and Severonikel were only two among many Soviet enterprises that were causing serious damage to the environment. The Soviet tendency was to dismiss the Western discourse as overly anxious.

More importantly, the modernization project lost steam because of the industrial and commercial interests invested in it. Originally, when the Soviet Union and Finland had talks about modernizing the Kola nickel industry in the Soviet–Finnish Economic Commission, the issue had been an industrial rather than an environmental matter. Outokumpu's early negotiations with the Soviet nickel industry had been firmly placed in a commercial and industrial context. This changed when the ecological realities in Pechenga became more evident to the Scandinavian countries in the late 1980s. The modernization concept turned into an environmental effort. Paradoxically, this did not remove the industrial or commercial hindrances, but exacerbated the situation. When Norwegian subsidization was announced in September 1990, and Elkem Technology came onto the playing field, another strong commercial interest had to be accommodated. Naturally enough, both Outokumpu and Elkem Technology were commercially motivated, and wanted to make the most of a potentially lucrative business deal. The price tag they put on the modernization project was prohibitive to a Soviet Union in deep recession. As we shall see in the next chapter, this dichotomy between industrial interests and environmental concerns was to persist when the Russian Federation took the place of the Soviet Union.

# Reconstruction time again

The 1990s were a decade of great inner turmoil in the Russian Federation. The political earthquakes that shattered the Soviet Union in 1991 brought a series of economic tsunamis that devastated the Russian social landscape again and again. Only towards the end of the decade did the re-born state show signs of picking up its imperial legacy and reasserting itself on the international political stage. Against the backdrop of the long lines of Russian and Soviet political history, the 1990s can be described as a 'state of emergency'.[1] Not since the middle of the ninth century, when the Slavic tribes along the Volga according to legend invited Scandinavian Vikings to rule them, had Russia lain so open to Western influence. This time, the Nordic neighbours were not invited to take over the realm – but they were certainly given chances to impact on developments in selected areas.

Many Western states seized the opportunity, investing substantial funds to alleviate the 'transitional problems' of Russia. This was not done solely for the benefit of the recipients but was equally motivated by the ambitions of the donor states themselves. A purveyor of gifts is also a wielder of power. And with money, one is entitled both to diagnose the patient and to prescribe the medicine.[2] In Norway, Russia's economic backwardness and social needs tended to fuel a discourse that was linguistically and conceptually akin to the language applied in foreign aid. Norwegian approaches to Russia became increasingly coloured by high-flying ambitions to reform this temporarily weakened, but nevertheless gigantic neighbour. In a sense, the historical asymmetry in the Russian–Norwegian relationship seemed inverted, at least to some Norwegian bureaucrats; Norway was, in their imagination, now taking

Workers leaving their shift at the Pechenganikel smelter, 1990s, while shrouded in gases emitted from the plant.
Photo: Ola Solvang

the lead. Whether they liked it or not, the Russian targets for Western aid in this period were in no position to say no. Here we may note how the pill was sweetened by a euphemism that had been pioneered in development aid. What were essentially aid programmes, in which Norwegian authorities transferred funds and expertise to Russia, were now referred to as 'cooperation' between neighbours.[3] Thus, although it was mainly the donor country that defined both the problems and their solutions, the outward appearance was one of unified action to solve shared problems. One arena for this type of camouflaged aid to Russia was the Russian–Norwegian environmental commission.[4] With the modernization of the Pechenganikel complex as one of its centrepieces, the commission soon developed a range of collaborative areas, all focused on solving post-Soviet environmental problems. Although the nickel modernization had been introduced when Russia was still part of the Soviet Union and had always depended on substantial financial input from the Soviet Union and briefly the RSFSR, its development in the 1990s was coloured by the general donor–recipient relationship between Norway and Russia in this period.

However, as the following discussion will show, any Norwegian/ Nordic expectations of a passively grateful Russia accepting generous gifts were doomed to disappointment. There might have been gratitude in evidence as regards less significant collaborative efforts, but in the case of Pechenganikel modernization, Russia's ability to withstand or ignore Nordic pressure proved strong. Rather than being forthcoming towards their affluent and environmentally minded neighbours, the Russian authorities and Norilsk Nikel turned the tables. In effect, the Nordic governments, and the Norwegian authorities in particular, were taken for a decade-long rollercoaster ride where expectations and high hopes were repeatedly crushed by disappointment and rejection. This pattern, apparent already in the early days of the Russian Federation, would become stronger towards the end of the decade.

Throughout the 1990s, the Pechenganikel project found itself squeezed and eventually stuck between environmental and commercial concerns, as we shall see. Norway's role in the matter soon outgrew the Nordic dimension of the initial phases, where Finnish and Swedish engagement had featured as central elements. Finally, we assess the development of the project, or lack thereof, from a higher vantage point. Both the bilateral Russian–Norwegian relationship and Russia-internal events are key factors to understanding the context, if not the fate, of the Pechenganikel modernization project. Therefore, the closing section of this chapter takes a closer look at Norway's overarching efforts to regionalize the European North through the establishment of the Barents cooperation, and offers a brief discussion of the privatization of Russian enterprises.

## Exit Outokumpu – enter PRC

The transition from having a Soviet counterpart in the modernization project, to facing a new Russian owner with a fresh though often confusing take on things, was soon felt in Norway and Finland. RSFSR deputy minister of the environment Liya Shelest, as we saw in Chapter 3, was quick to cast doubts about the future of the modernization project, stating that Pechenganikel might be shut down before long, which would render costly investments superfluous. Clearly, the Russian Federation found the Finnish–Norwegian project overly expensive and wanted to look into cheaper solutions.[5] However, reports of diminishing ore resources at Pechenganikel were accompanied by rumours in a completely different direction. One example was the talk of Pechenganikel's plans to expand its mining operations with the help of Canadian technology, which would allegedly enable continued activity with the local ore deposits for another fifty years.[6] When Outokumpu representatives visited colleagues in Nikel in early January 1992, they were shocked to hear that Russian authorities purportedly no longer saw any need for modernization at all. According to the Pechenganikel management, the new Russian leadership referred to a 1980 agreement with Finland providing for 50 per cent cuts in emissions from the 1980 level (see Chapter 2), and now stated that this target had been almost achieved: full compliance was to be reached through the introduction of new (Russian) technology.[7] Modernization with Western help at Pechenganikel, it seemed, was unnecessary.

The project was not abandoned, however. Mixed messages coming from the Russian side in this period did not mean an end to the modernization story, but instead infused it with a new dynamic. While the Soviet response to Nordic pressures and plans had been defensive, consisting mainly of reiterating that the modernization project was simply too costly, the Russian responses were more aggressive. Although they were still in a precarious economic position, Russian authorities and the Norilsk Nikel state concern were examining alternative upgrades for Pechenganikel and discussing options with the Canadian company Falconbridge. This unsettled Outokumpu. Elkem Technology, by contrast, had been contacted by Falconbridge on this matter, and decided to hedge their bets by staying open to possible future collaboration with the Canadian giant, should Outokumpu's proposal fall through. In this, Elkem was supported by the NME.[8] Most importantly, though, the Russians had changed the framework of the modernization project: they would not be forced into any agreements with Finland or Norway or both, but wanted to keep all options open, even if that might entail the loss of Nordic funding.

It is likely that Russians were aware not only of the strong interest in the modernization project among Norwegian industrialists, but also of the prestige that the Norwegian government had already invested in the project.[9] Thus, any risk of losing access to Norwegian funding must have seemed negligible. Norway's persistence in the matter was underscored during three official visits in 1992: one by Russian MPs to Norway, one by the Norwegian minister of foreign affairs to Moscow and one by his colleague Andrei Kozyrev to Oslo. On all three occasions, Norwegian representatives called for a speedy Russian decision on technology, while also signalling that other solutions than the one Outokumpu and Elkem Technology had presented in late October 1991 might be considered for Norwegian funding.[10] The Russians were finding that they had plenty of room for manoeuvre.

During the first part of 1992, the Russians seem to have been exploring the boundaries in this manoeuvring room. While Outokumpu was still awaiting a decision regarding their proposed project, and the 31 March deadline was nearing, mixed messages kept coming from Moscow. In talks with representatives of the Norilsk Nikel state concern, Outokumpu was told that everything was set for technical negotiations – meaning that the Outokumpu tender had in reality been accepted.[11] However, nothing was heard from Norilsk Nikel until June, when President Yeltsin intervened and decreed that modernization negotiations must move forward.[12] In the meantime, Outokumpu had already moved the deadline to the end of October 1992.[13]

This demonstration of Finnish (and in effect Norwegian) eagerness and malleability may well have served to encourage the Russian Ministry of Economic Affairs, which had been temporarily commissioned with responsibility for Russian management of the modernization project. When ministry representatives met in Moscow with Finnish and Norwegian representatives on 3 June, Deputy Minister Yurii Olkhovikov declared that Russia was prepared to commence with the modernization, in full accordance with Outokumpu's tender. However, he added, there was one major condition that would have to be met. To the general amazement of the Nordic partners, he now demanded that their combined aid package be increased from USD 100 million to USD 450 million, which would reduce Russia's portion of the costs to a manageable USD 190 million.[14]

The real intentions behind Olkhovikov's rather surreal demands would emerge shortly. Predictably, the Finnish and Norwegian representatives immediately declined the proposal – a refusal that could hardly have come as a surprise to the Russian side. The Russians did not budge, however. Olkhovikov, Russian minister of the environment Viktor Danilov-Danilyan and even Boris Yeltsin stood firm in demanding that the Norwegian and Finnish financial support be increased to cover at least 70 per cent of the total project costs. Three months later, when the Nordic and Russian Ministers of the Environment met in Kirkenes on September 3, the question was brought up one last time. While Danilov-Danilyan had no choice but to reiterate the Russian position, he also took a new tack. Well aware that his Nordic colleagues could not accept the Russian conditions, he proposed they all commence a search for a new and improved – and above all, less costly – project for modernizing the Pechenganikel metallurgical plant.[15]

In effect, this meant the end to the tender from Outokumpu and Elkem Technology. Their solution had met with what was in fact a predictable fate. At a cost of USD 640 million (the last USD 40 million USD had been added to cover a sports complex and a health facility), it was simply too expensive. By posing impossible demands, the Russians had simply made sure that this was clear to both sides of the table.

Viktor Danilov-Danilyan's call for renewed efforts to find a financially and environmentally acceptable solution to the sulphur dioxide problems of Pechenga was well received, at least in the NME. In fact, Norwegian civil servants seemed to think that the new turn of events made for better prospects and underscored that nothing was changed with regard to their funding. Provided that a scheme to secure a 90 per cent cut in sulphur dioxide and heavy metals discharges from Pechenganikel was tabled, Norwegian funding would follow. Even Sweden, which had played a peripheral role, now intimated that Swedish financial support would be possible.[16] Although the rejection of Outokumpu's tender must have been a blow, Finland was still interested in a clean-up on the Kola Peninsula. In October, at the behest of the Nordic and Russian Ministers of the Environment, a group of experts met in Oslo to consider alternative and less costly ways of curbing the discharges. Their findings were presented to the ministers of environment when they met again in Copenhagen in late November.[17]

In Copenhagen, the Russians yet again changed the game, however. Danilov-Danilyan now announced that instead of inviting selected manufacturing firms (presumably from the donor countries) to develop alternative projects, Russia preferred to have the modernization put out to tender, inviting all interested companies to outline their solutions.[18] With this, the new Russia took yet another step into the world of international capitalism, claiming the right to choose freely from a range of offers. Indeed, the Russians also reserved the right to engage domestic industry in the project, to bring costs in foreign currency down to a minimum.[19]

The tender deadline was eventually set to 15 November 1993. In fact, this approach did not change the outcome significantly. Elkem Technology, having foreseen that the Outokumpu project in which it was involved would be financially unacceptable to the Russians, quickly activated an alternative solution. In partnership with Swedish Boliden Contech and Norwegian Kvaerner Engineering, Elkem Technology submitted a new project for consideration. The offer from the three companies, now operating as the PRC, aimed at a 90 per cent reduction in emissions at a cost of USD 300 million – half the price that had been quoted by Outokumpu. On 6 January 1994, the new consortium was informed that their outlined project was preferred by the Russian tender committee, and the three companies were invited by the Norilsk Nikel state concern for further negotiations in St. Petersburg the same month.[20]

As would soon become apparent, this by no means meant that work could now commence in Pechenga. Quite the contrary – many factors still needed clarification, and long negotiations lay ahead before the Russian government would approve of the plans. There was still the financial question to be sorted out; and PRC's winning tender had been only loosely developed on the technological side. In fact, that had been necessary, as many facets of Pechenganikel's future production were unclear and undecided. Most importantly, the Norilsk Nikel state concern had no firm plans regarding the quantity of Norilsk ore that should supply the Pechenganikel smelters: the state concern had been mooting the possibilities of refining anything from substantial amounts of their Siberian ore to solely local ore from Pechenga. This question was of vital significance, as the sulphur-heavy Norilsk ore would call for different technological solutions than what was required for the local ore.

One sceptic, who had been involved in developing alternative reconstruction projects after the Outokumpu proposal had been rejected, warned the NME: discussions between the Nordic companies and Norilsk Nikel were in fact a mock process; the Russians were actively avoiding clarification of the ore question and other issues. He wrote: 'Unfortunately it can be surmised that the Russians are completely aware of the loose foundation of the whole endeavor, but still choose to carry on as this demonstrates their good will in a project that they at the present time see no economic advantages in.'[21] The warning of this whistleblower was not heeded at the time. However, considering subsequent events there is much to suggest that he was on to something.

## Nickels and dimes III

As in the previous round, the host of pecuniary uncertainties related to the project was proving problematic. One thing was the overall funding of costs estimated to anywhere between USD 200 million and USD 300 million, depending on Russian decisions as to which ore to use.[22] Although halving the costs of the Outokumpu project helped, it would still be difficult to raise enough funds. To encourage generosity among Nordic governments, a direct link between access to participation in the project for potential subcontractors and funding from their national authorities was communicated, by the private companies involved and the Norwegian authorities alike. For example, the PRC was adamant that Finnish subcontractors would be chosen only if the Finnish state contributed financially, most notably in the form of export credits, to the completion of the project.[23]

The Finnish and Swedish governments proved unwilling to commit themselves. To apply moral pressure, Norwegian minister of the environment Thorbjørn Berntsen approached his Swedish and Finnish colleagues directly. Referring to their shared interest in curtailing emissions from the Kola Peninsula and reminding them that industrial interests in both countries were potential benefactors from the modernization endeavour, he asked for contributions.[24] These efforts appear to have been in vain, however. After the Outokumpu project had been rejected, Finnish interest cooled substantially.[25] The Swedish government was increasingly reluctant to continue funding Boliden Contech's project planning after the PRC tender had been accepted, arguing that the consortium should be able to stand on its own feet from then on.[26]

For the industrialists, the accumulating project planning costs quickly became a source of trepidation. The PRC companies had devoted many hours to negotiations and further development of the modernization scheme, and as yet no actual construction work had begun. In their view, this effort would have to be compensated for by the Norwegian and Swedish governments, as they deemed the economic and political risks of the project to be still very high. When the Swedish government seemed unwilling to offer support, Elkem Technology included Swedish Boliden Contech's costs in their funding applications to the NME. According to Elkem Technology, Boliden Contech's participation was crucial to the future of the project. Knowing that the Swedish company would withdraw from further preparatory work if expenses were not covered, Elkem Technology claimed it was essential for the Norwegian authorities to step up and open the public wallet even wider.[27]

Risk aversion in the Nordic consortium was pronounced. The three companies were clearly unwilling to jeopardize their own funds in planning a project for a buyer as unpredictable as a Russian state concern. On the other hand, given the readiness of the Norwegian authorities to spend public money, project development itself was good business for the consortium. For a relatively brief period of project planning and negotiations during the spring of 1994, the three companies in PRC budgeted for expenditures just short of USD 1.4 million.[28] The NME agreed to cover 75 per cent of this, and the consortium companies were able to seek reimbursement of their 25 per cent share if the project should fall through.[29] Although this was in part a loan (*tilbudsgaranti*) – the money was to be paid back in full if the modernization project did materialize – in reality it protected the PRC against any losses. In light of the scale of this endeavour (PRC designated 26.5 fulltime positions for planning during the spring 1994),[30] it is safe to say that the planning work alone, and the sums paid out for it, were proving sufficiently lucrative to foster enthusiasm among the companies involved.

Finally, on 5 December 1994, PRC was able to present an adjusted offer to the Norilsk Nikel state concern at a reduced cost. For USD 257 million, the Pechenganikel combine would be thoroughly revamped and emissions reduced by 90 per cent from the 1980 level. The offer would stand till 31 March 1995.[31] Again, however, there was trouble ahead. Although costs had been cut to less than half of Outokumpu's price, there still was little to indicate that a crisis-ridden Russia would be willing or indeed able to raise in excess of USD 200 million for an environmental project. Amidst all this, the already doubtful Finnish and Swedish commitment to the modernization was becoming increasingly uncertain. One centrally placed Norwegian civil servant gave the following analysis (formulated in tortuous bureaucratic language) of the Norwegian and Russian positions:

> The background for the whole issue seems to be a strong Norwegian wish that the Russians clean up their smelting plant and that they [the Russians] should be willing to accrue debt to achieve this. The Russians, for their part, seem to have clearly expressed that Norway,

being an affluent state, has such a strong self interest in this project that unless a help package significant enough to be deemed sufficient by the Russians is put together, the project would be of little interest to Russia. One might think that the Russians this time will want to present themselves with a unified voice and give clear signals, but it is still hard to free oneself from the thought that unless the Norwegian/ Nordic gift package is brought up to a level substantially higher than earlier, there will be little interest in the project on the Russian side. There is at this time therefore a pressing need for the Norwegian side to meet with Nordic partners to agree how much they are willing to contribute[.]

This civil servant, however, did not see much hope for Finnish and Swedish commitment to funding:

As this project will mainly be to Norway's gain, there is little to suggest that Finland or Sweden will show much interest in it. . . . Previously, we have been informed that Finland, on aggregate level, has USD 10 million at their disposal [for use in the post-Soviet area], but it is doubtful that much of this will be used [for Pechenganikel modernization] unless significant Finnish subcontracts are made in the project. On the Swedish side there is very limited interest in the project as problems in the Baltic Sea naturally preoccupy our Swedish neighbors.[32]

Thus, there seemed to be scant hope of raising Nordic funding for the modernization. Prospects were even bleaker further east. The deteriorating economic situation in Russia at the time and the very problematic fiscal development within Norilsk Nikel in the fall of 1994 gave little reason for optimism.

Nevertheless, to sway Russian decision makers before the PRC deadline expired at the end of March 1995, the Norwegian minister of the environment held several meetings in Moscow in late February. In talks with the management of Norilsk Nikel, the involved Russian ministries and Deputy Prime Minister Aleksandr Zaveryukha, Berntsen and his entourage sought to convince their Russian hosts of the need to complete the modernization. One important message was that the project would, contrary to widespread depictions, actually be profitable in the long run. The Norwegians pointed out that the long-neglected need for maintenance at Pechenganikel would be taken care of, and that the project would result in more efficient ore exploitation and lower energy consumption.[33]

This message, emphasizing economic rather than environmental benefits, met with varied Russian reactions. One situation, recorded in the minutes of the Norwegian entourage's meeting with a delegation headed by Deputy Minister of the Economy Vladimir Kossov, may illustrate the inner tensions among the Russians:

> Kossov stated that irrespective of how the project is funded, it will always entail a cut in [the Russian] tax base and a reduction of tax income. Consequently, the burden will in any case be put on the Russian taxpayer's shoulders. He pointed out that the budget deficit has reached 73 trillion rubles (or about USD 20 billion). The company Norilsk Nikel will have their problems solved while the taxpayers hurt [Kossov said]. Director [of Pechenganikel] Igor Blatov took the floor. He disagreed. Pechenganikel is the world's fifth largest producer of nickel. After having invited people to settle in the North, the state has now disclaimed the responsibility of keeping residents in the area and social benefits have become the responsibility of the enterprise. All the 38 thousand tons Pechenganikel produces today are sold abroad. With an export tax of USD 800 per ton, this gives state coffers USD 30 million annually. If you add export of copper and other metals it is even more [Blatov said]. By this point, Kossov and Blatov were bickering. Kossov challenged Blatov to say whether his enterprise is in fact paying [taxes] or not. Blatov said that the state has always been paid its due. The current funding scheme entails that 15% is covered by Norway, 15% by the Russian state and 70% by Norilsk Nikel [Blatov added]. Kossov brushed this off and said that it was always easy to seek funding 'from another man's pocket'.[34]

Kossov and Blatov's very public row, apart from demonstrating the lack of Russian unity in this matter, was a reminder of the many crossing conflict-lines in post-Soviet Russia in the mid-1990s.[35] The ongoing privatization of previously state-owned enterprises was a complicated matter, especially for large employers like Pechenganikel.[36] Many fractious points cropped up, not least the division of responsibilities for the welfare of the employees/citizens between the state and the companies that were being denationalized. Importantly for the modernization project, this did not bode well as regards the Russian ability to make decisions.

The main problem was, of course, how to divide the Russian modernization expenditures. Blatov's claim that Pechenganikel would in reality account for about 70 per cent of the costs referred to the fact, although the company had been gradually privatized since 1994, that the Russian state still controlled most of its revenues. Income taxes, export taxes and company debt owed to the state assured this. For Pechenganikel to be able to contribute to the modernization, then, the state would have to be willing to grant the company a tax break, at least temporarily. As we see from Kossov's comments above, he was reluctant to accept such an arrangement. Pechenganikel's substantial contribution to state revenues was clearly too important to be forfeited, especially if the purpose was to alleviate environmental concerns that were trumpeted primarily by a neighbouring state. The NME concluded upon the return from Moscow that, although the industrialists in Norilsk Nikel and especially Pechenganikel seemed genuinely interested, the crux of the matter would be the question of tax breaks for Pechenganikel. It was clear to the Norwegians that this problem would not be solved if left only to lower-level actors.[37]

Through February and early March, the usual mixed messages kept coming from the east. The Russian actors who stood closest to their Norwegian colleagues, like the Minpriroda official Larisa Yanchik and her superior Danilov-Danilyan, expressed great hopes for the project while referring to good progress in negotiations between Norilsk Nikel and the relevant ministries.[38] A more sober assessment, albeit still optimistic, came from Pechenganikel director Igor Blatov. While hopeful of a positive outcome, Blatov agreed with the Norwegians and insisted that it was necessary to involve the topmost echelons of the

Russian political elite. Here he made specific mention of President Boris Yeltsin and Minister of Foreign Affairs Andrei Kozyrev, the latter representing Murmansk oblast in the State Duma. Only with their help, Blatov stated, could the red tape be cut through.[39]

Getting a straight answer from Russian high-level politicians in the Pechenganikel issue was easier said than done, however. By the mid-1990s, the Russian taxation system, which since the breakdown of the Soviet Union had been arbitrarily enforced,[40] had become highly controversial. Overall, Russian tax authorities managed to collect only a fraction of their dues. A tax break to a huge company like Norilsk Nikel, which was rumoured to have masses of hard currency stacked away in foreign bank accounts,[41] would fuel further criticism of an increasingly lacklustre government.

In a bid to regain some credibility in tax matters, President Yeltsin issued a series of decrees to tighten up the tax and tariffs regime. In March 1995, Minister of the Economy Evgenii Yasin thus found himself in a difficult position when asked to support a drastic reduction in the export duties of Norilsk Nikel. Although he did support the Pechenganikel modernization funding scheme, he was reluctant to condone it publicly, after having recently associated himself with Yeltsin's hardline stance in tax questions. Instead, in a deeply ambivalent compromise, Yasin left it to a lower-ranking member of his ministry to support the proposal. In effect, then, the scheme was formally supported by the ministry as such, but not by its leader.[42] This non-decision on the part of Yasin deepens the impression of a Russian government hamstrung by internal division. The chaotic and factious nature of Russian politics, with cross-cutting interests and allegiances all the way through to the presidential administration, made for inefficient governance. Prospects of a decision on the Pechenganikel modernization project seemed bleak – the Russians did not seem sufficiently unified even to say 'no'.

The Norwegian government, intent on pressing for a Russian decision on funding, brought out its big gun. Prior to her visit to Moscow in May 1995 (to celebrate the fiftieth anniversary of the end of the Second World War in Europe), Prime Minister Gro Harlem Brundtland sent a letter to her colleague Viktor Chernomyrdin, stressing the importance of a speedy conclusion to the question of Russian funding.[43] This, however, did not seem to have the desired effect. When Brundtland brought the question up during her visit to Moscow, neither Chernomyrdin nor Deputy Prime Minister Oleg Soskovets committed to the funding scheme developed by Norilsk Nikel. In their view, a company of Norilsk Nikel's stature (Norilsk Nikel was quite prosperous by Russian standards – also in the mid-1990s) should be able to bankroll the project on its own, especially given the Norwegian contribution. Both were, at best, willing to offer support by granting Norilsk Nikel favourable terms on government loans for investments.[44]

The Russian political setting that met the Norwegian prime minister was extremely fragmented. On the one hand, there were two ministries – economy and environment – that both supported Norilsk Nikel's appeal for reduced export duties. This support was conditional, however, inasmuch as the minister of the economy himself had strong reservations about publicly backing the scheme. Furthermore, the same government's prime minister and his deputy, as well as the Ministry of Finance, were opposed to the funding arrangement. They thought that the project ought to proceed only if Norilsk Nikel agreed to assume more of the financial burden.

Matters were further complicated by the fact that the modernization project was caught up in the ongoing privatization of Norilsk Nikel. As a result of the highly complicated denationalization process of state enterprises, which entailed not only a core restructuring of the production system but also massive reforms in what little was left of the post-Soviet welfare services, Russian politicians were eager to point out that the wellbeing of former state enterprises was no longer their responsibility.

Despite the many conflict-lines, the Russian government did issue a decree dated 1 July 1995, promising support for the project. With reference to the Convention on Long-Range Transboundary Air Pollution (see Chapter 2) and considering ambitions to improve environmental conditions on the Kola Peninsula, the Russians declared willingness to match Norwegian funding and to further assist by channelling an undisclosed figure in foreign currency accrued from Pechenganikel's export income. The decree specifically charged the Russian Ministry of Foreign Affairs, the Russian Ministry of Foreign Trade and the Russian Ministry of the Environment with responsibility for attracting contributions from Finland, Sweden and other countries to top up the project funding.[45]

Although this decree was an unprecedented expression of Russian willingness to contribute to the modernization, it hardly paved the way for the project to start. Even with Russian matching of the Norwegian gift package, more than three-fifths of the price tag of USD 257 million remained unaccounted for. Loose formulations about transfer of future currency revenues from nickel exports were a weak foundation on which to start a massive modernization project. Pechenganikel director Igor Blatov and technical director Gulevich in Norilsk Nikel both found the decree wanting. If the Russian government in fact did decrease or remove export duties, it would, they claimed, instead increase Norilsk Nikel's tax burden to make up for lost revenues. In their view, the decree did little to clarify the composition of Russia's contribution.[46] That said, given the ever-changing Russian economy and political scene, this was probably the best the Norwegians and Norilsk Nikel could have hoped for at the time.

## PRC falters – negotiations stall

The NME was at this point becoming acutely aware of the wobbliness of Russian statements concerning the modernization of Pechenganikel. Although the ministry had reservations about the Russian government decree, they at least saw it as a positive sign and were eager not to rock the boat. Then, their trepidations rose to unprecedented heights when Moscow's unpredictability seemed to have rubbed off on PRC. In the summer of 1995, Elkem Technology – to everyone's surprise – seemed intent on withdrawing from the consortium.[47] When the time-limited Norwegian–Swedish partnership formally expired at the end of June, Elkem Technology did not renew the agreement. The company argued that repeated Russian breaches of deadlines (31 March 1995 and 30 June 1995) and costs related to the prolonged upkeep of a large staff for a project fraught with risk factors had forced a reassessment of the company's position. Doubtless, the many mixed messages from Russian players and their apparent disregard for agreed deadlines had created considerable irritation in Elkem. The company director said he was left with the feeling that while the Norwegian industrialists were committed to deadlines, the Russians seemed completely unbound by all agreements.[48]

In effect, Elkem Technology's refusal to renew the consortium agreement meant that PRC ceased to exist as a legal entity.[49] Worried that this would unsettle the Russians and give them a pretext for further delays, the Norwegian minister of the environment implored Elkem Technology to remain committed, at least outwardly.[50] The company agreed to continue as a nominal partner, and to participate in further negotiations with Norilsk Nikel and Pechenganikel. Furthermore, Elkem Technology declared that they would honour all delivery contracts if the modernization project did in fact commence.[51]

However, by not prolonging the consortium agreement, Elkem Technology gave a strong signal of discontent with the lack of progress in the project.[52] Elkem Technology's demonstration illustrated one of the central disparities between the Scandinavian consortium and its Russian customer: the value put on *time*. While the PRC wanted a binding contract as soon as possible, Norilsk Nikel felt that lengthy talks were of the essence. In August 1995, just before the consortium and Norilsk Nikel/Pechenganikel were to embark on yet another round of contract negotiations in Zapolyarnyi, Norilsk Nikel's technical director Gulevich commented on the time frame to the Norwegian embassy:

> Gulevich was of the opinion that contract negotiations would [have to] be time consuming. The contract in question was wide-ranging

and expensive and the company [Norilsk Nikel] would meticulously examine all its aspects. . . . The consortium [PRC], however, pushed for a speedy closure. It seems that Gulevich finds this pressure somewhat bothersome.[53]

The Scandinavian industrial actors were at this point, in mid-1995, beginning to turn rather glum. The energy that had infused the PRC at its inception in early 1993 had gradually been sapped by repeated last-minute Russian decisions that for various reasons delayed the project. To make matters worse, the Russians were calling for even more patience. For Elkem Technology, frustration with their foot-dragging Russian counterpart led the company to the verge of withdrawal. Ironically, this was barely known to the Russians. Both the NME and the prime minister's staff agreed that the news of Elkem Technology's diminishing role in PRC should be communicated with a minimum of fuss. One of the prime minister's aides mentioned it in passing to the Russian ambassador to Norway, while assuring the ambassador that it would have no consequences for Elkem Technology's commitments.[54]

Also on the governmental level, Norway's patience was being tested. In May 1995, when Gro Harlem Brundtland visited Moscow, she agreed with Deputy Prime Minister Oleg Soskovets that a 'memorandum of understanding' between the two governments should be set up to bolster the project.[55] Basically, the document was intended as a brief text where the two states declared that they would stand by their commitments to fund the modernization of Pechenganikel. Even before Viktor Chernomyrdin issued his decree on Russian funding (1 July 1995) an agreement text was being prepared in Norway.[56] In late June, a month before a planned visit by Yeltsin to Oslo, the Norwegian government was informed that the text had been approved in Moscow.[57] The Ministers of Foreign Affairs, Andrei Kozyrev and Bjørn Tore Godal, were expected to sign the document during Yeltsin's visit. However, the presidential trip was postponed. At the behest of the Russian ambassador to Norway, the Norwegians prepared for signing the next time the two ministers of foreign affairs would meet, in London on 21 July that year.[58] This was not to be, however. One day before the London meeting, the Norwegian embassy in Moscow was informed that the Russian government could not accept one of the articles in the memorandum and demanded its removal.[59]

The problem was Article 4, which specifically addressed the need for Russian state guarantees. The Russians argued that their responsibility to assure project completion was sufficiently described in Article 3 of the memorandum, which stated that the Russian government was expected to make available all funding that was not covered by the Norwegian contribution. Presumably, the disputed Article 4 was based in Norwegian fears that the project would come to a halt sometime after the full Norwegian contribution had been paid out. The Norwegian Ministry of Foreign Affairs envisaged a situation where the Russians would simply opt out, leaving it to Norwegian authorities to fund the completion of the project.[60] In the end, it was those in the Ministry of the Environment, who had invested the most work, prestige and pride in the Pechenganikel modernization project, who made an almost emotional plea for their colleagues to accept the Russian demands to remove Article 4:

> The memorandum was initially politically motivated. It was never intended to create a legally binding agreement that covered all possible events. If the Russian government opts to carry the political burden associated with breaching the agreement, the inclusion or non-inclusion of article 4 would hardly matter. Doing business with today's Russia in projects of this magnitude requires a readiness to replace some of the usual legal bindings, guarantees and 100% security with trust.[61]

This plea was, despite all its un-bureaucratic naivety, in a sense quite realistic. There was, considering the lack of Russian urgency in the modernization process, little to suggest that the Kremlin would agree to terms that did not suit them. The plea was probably also a result of decreasing patience among Norwegian environmental bureaucrats. Within the Ministry of the Environment, there seems to have been a sense that the modernization project had reached a decisive moment: It was now or never. After almost five years of painstaking negotiations on matters both large and trifling, efforts at last stood a chance of coming to fruition. Too many Norwegian demands and objections could estrange the Russians and thus render all the work useless. In the end, this fear carried most weight in the question of Article 4. Norway's decision to agree to its removal was communicated to the Russian ambassador to Norway on August 25.[62] The agreement itself was signed by the two foreign ministers on 9 September 1995.[63]

Contract negotiations between PRC and Norilsk Nikel/Pechenganikel were intended to take place on two levels. Engineers from both sides were tasked with finalizing all technical details in separate talks. Parallel negotiations on the executive level were meant to secure consensus on all pending questions related to funding and payment mechanisms. This was of course contingent on the final settlement as to how the Russian funding would be built up.[64] The technical negotiations started in mid-August in Zapolyarnyi. To the discontent of the Norwegian side, they promised to be just as time-consuming as Gulevich in Norilsk Nikel had foreseen. PRC's project manager reported back to Oslo that the thirty-member Russian delegation in Zapolyarnyi was disorganized and on many points badly informed about the content of the modernization project.[65] Norilsk Nikel further frustrated the Norwegians by requesting additional technical assessments of the modernization plans. There seemed little hope of a settlement before mid-November[66] – but the technical negotiations were at least proceeding.

Executive-level negotiations on funding mechanisms were put on hold pending a firm Russian government decision on financial support. In late October, reports from Moscow indicated that efforts to secure Russian funding had encountered problems. The matching of the Norwegian contribution that was promised in Chernomyrdin's July decree did not appear in the Russian national budget for 1996, and the Russian Ministry of Finance refused to push for removal of duties on Norilsk Nikel's exports. The Norilsk Nikel management, for its part, did not wish to conduct contract talks with PRC before the funding issues were settled.[67] Thus, instead of entering a period of intense deliberations, the industrial actors were obliged to spend the rest of 1995 waiting for action to be taken by Russian authorities. For the Norwegian onlookers, who had originally hoped to conclude negotiations by late summer/ early fall 1995, this lack of progress was highly frustrating. Several letters were sent, both by the industrial actors and by the Ministry of the Environment, to their Russian counterparts, requesting a speedy resolution of the problems.[68] There was little or no response from the Russian side.

As the end of 1995 approached, the technical negotiations that had been going on in Zapolyarnyi since August were still underway. Information given to the NME indicated that the two sides had difficulties in finding a common language:

> The reports from the negotiations indicate a somewhat stiff and perhaps not very creative negotiation atmosphere, and that there might be a need for some outside intervention. It also appears that the management of Pechenganikel is unclear regarding their future production, that they do not trust their own ability to solve technical problems and that they might have unrealistic expectations about what they can achieve through these negotiations.[69]

A new and unexpected (to PRC) issue in the negotiations concerned the mix of raw materials to be processed in Nikel. Having agreed with PRC in spring 1994 that only local ore would be smelted in Pechenga, the Russian negotiators now required that the modernized process be developed to be able to handle up to 300,000 tons annually of the sulphur-heavy Norilsk ore. This had wide-ranging consequences for both the scope and the price of the modernization. Above all, it would make it much more difficult to achieve the discharge limits that had been defined as part of the Norwegian gift package.[70]

The Russian unpredictability probably reflected the volatile privatization process underway in the Norilsk Nikel concern at the time. Long-term planning for Pechenganikel's future was difficult as long as the ownership situation remained undecided. It also attested to what had been observed by the Norwegian delegation at an early stage in the Zapolyarnyi talks– that many of the Russian delegation members were ill-prepared and unfamiliar with the actual content of the modernization project and what had been agreed upon before they became involved.[71] This led to ineffective negotiations and gradually wore down the Nordic partners in PRC.

## Procrastination and dissolution

PRC had problems of its own. No longer was it a harmonious union of Norwegian and Swedish industrialists. After the consortium agreement expired and Elkem Technology partly withdrew from PRC in the summer of 1995, the lack of progress in the modernization project was seriously undermining the already weakened Norwegian–Swedish pact. Kværner Engineering, now representing the formally non-existent and slowly dissolving consortium, had strong reservations about the Swedish and Norwegian partners. Of Elkem Technology, the project manager said that they had been less than cooperative. Of the Swedish partner Boliden Contech, he made it clear that their capacity and competence were considered unsatisfactory. Nonetheless, Kværner Engineering was willing to continue the cooperation.[72]

It was more problematic, then, that the many delays on the Russian side stretched the resources of the consortium – primarily those of Kværner Engineering. Commercially speaking, it was highly questionable to maintain activity in a project that seemed to be going nowhere. Project staff set up specifically for the Pechenganikel modernization had to be relocated to active duty elsewhere.[73] In January 1996, Kværner Engineering downscaled the staff to a minimum.[74] With no movement on the Russian side, modernization was put on hold, and a certain project fatigue was palpable among the industrial interests and Norwegian environmental bureaucrats and politicians alike.

New developments temporarily changed this, however. Modernization talks were revived when President Yeltsin, after several cancellations due to recurrent heart problems, finally visited Norway in March 1996. That was a good time for a visit. With presidential elections approaching, the faltering Yeltsin was keen to demonstrate his qualities as a world leader who was respected and popular in other countries. The visit would be broadcasted to the Russian electorate, and Yeltsin's need to win over the people made for constructive negotiations in Oslo.[75] For the Norwegians, the result was encouraging. An agreement between Norway and Russia was signed by Norwegian minister of the environment Thorbjørn Berntsen and Russian Deputy Prime Minister Oleg Soskovets. The brief text was seen as an important breakthrough. It stated that the Russian government would exempt Pechenganikel from all export duties (thus enabling the enterprise to pay for its share of the costs of modernization) and imports of modernization equipment and installations from import duties and value added taxes (thus lowering the costs).[76] For good measure, the boisterous and impulsive Russian president gave his own government members a dressing-down for their inertia in the modernization efforts so far – in front of the Norwegian business community and the international press.[77] The Russian media, controlled by oligarchs who favoured Yeltsin for business reasons, praised the jovial Yeltsin for his positive visit to Norway.[78]

Although Norwegian environmental bureaucrats and representatives of Kværner Engineering were, according to one highly reputed Norwegian newspaper, aglow with optimism after Yeltsin's visit,[79] their hopes of a speedy start to the modernization project proved to be a pie in the sky. The following months would amply illustrate the limitations to Yeltsin's presidential powers. True, he was able to provide the necessary customs exemptions and tax breaks, which left the privatized Norilsk Nikel with no excuses for not going ahead with the project. As events would show, however, what an increasingly toothless Yeltsin stated on a trip abroad meant little or nothing to those controlling both large enterprises (including Norilsk Nikel) and parts of the government apparatus in Russia from the mid-1990s. The oligarchs, having rescued Yeltsin from a humiliating defeat in the 1996 elections, were now claiming their dues.

Not long after Yeltsin's visit, negotiations between Pechenganikel and Kværner Engineering hit yet another snag. News of the silent dissolution of PRC had finally reached Russian ears, and, as predicted in Norway, it did create some problems. In a letter to Kværner Engineering's top management, Pechenganikel director General Igor Blatov vehemently protested against what he understood as the company's policy of 'pushing the leading technology enterprises Elkem Technology and Boliden Comtech [sic] out of the consortium'.[80] Blatov's conspiratorial interpretation of Kværner Engineering's motives is symptomatic of the difficulties the industrial parties had in communicating with each other. The reality, namely that Elkem Technology in mid-1995 had decided to tone down its engagement in the modernization primarily due to Russian inability to move the modernization forward, seems to have been unknown to the Pechenganikel director.

Blatov's letter gave rise to consternation in Oslo, and Thorbjørn Berntsen wasted little time in trying to bring matters in order. In a letter to his colleague Viktor Danilov-Danilyan, he assured the Russians that both Elkem Technology and Boliden Contech would still be available for the modernization project even though PRC was formally dissolved. The Norwegian minister further urged Danilov-Danilyan to inform both industrial and political actors in Russia about this.[81] Berntsen's letter was in Oslo thought to have had the desired effect. Reportedly, Pechenganikel was instructed by Danilov-Danilyan to go through with the negotiations.[82]

As minister of the environment, Danilov-Danilyan had only limited influence over one of Russia's major industrial enterprises, however, and Norilsk Nikel remained reluctant to commit to the modernization. In negotiations in Zapolyarnyi in June, Blatov claimed that his employer had no money for project funding, despite the lifting of export duties, and requested Norwegian funding to get the modernization going.[83] This request came as a surprise to the Norwegians.[84] After all, allowing for Pechenganikel investments in the modernization had been the rationale for giving Pechenganikel dispensation from Russian export duties, a victory so hard-won only a few months earlier. When he confronted the Russian authorities with this, the Norwegian ambassador received assurances as to Russia's commitment to the modernization.[85] Kværner Engineering was given similar reassurances in mid-summer 1996 in meetings with the newly appointed chairman of the board of Norilsk Nikel, Aleksandr Khloponin.[86] However, when there was still little or no movement on the Russian side by September 1996, Kværner Engineering and the NME agreed that the project staff should be cut down to a minimum, pending new developments.[87]

On 21 October 1996, Norwegian authorities informed the Russians that they had until 11 November to decide whether to commit to the modernization project or not. The message was that neither Norwegian authorities nor the companies involved could keep project preparations going indefinitely, and that Norway would have to retract all financial support in case of further delays.[88] The fate of the modernization seemed sealed when Kværner Engineering in late October were informed that Khloponin, contrary to his own reassurances only a few months earlier, had now decided to shelve the project altogether.[89] In mid-November, having heard nothing from Norilsk Nikel, the Norwegian Ministry of Foreign Affairs developed plans for withdrawal of all Norwegian support for the modernization scheme. The main goal was to avoid a situation where the Russian could 'place the responsibility for the expiration of the agreement [to modernize Pechenganikel] on Norway'. It was essential to present the Norwegian decision in a manner that would avoid this.

Surely, then, one would think that the modernization project had finally reached its end point. The Russian industrialists were not entering into the necessary agreements with Kværner Engineering despite pressure from their own authorities. Even the holder of the highest office in Russia, Boris Yeltsin, did not seem to carry enough clout to decide the matter. And yet, conflicting signals kept coming. On 6 November 1996 – a mere five days before the deadline set by Thorbjørn Berntsen – the managing director of Pechenganikel, Igor Blatov, urged Kværner Engineering not to give up on the modernization, and mentioned a range of factors that had halted contract negotiations.

First of all, he explained that the lifting of customs duties had lost its effect after the Russian state had introduced other taxes to make up for the export duty losses, just as he had predicted. Moreover, the rising costs of energy weighed heavily on Pechenganikel, and the enterprise was in no position to make investments. Indeed, the situation was so bad that Pechenganikel had for a while barely been able to pay salaries to its workers. Blatov also said that the unsettled political situation after the elections in June, exacerbated by President Yeltsin's continued cardiac problems, had contributed to the delays.

In the short term, Blatov's last-minute appeal seems to have had the desired effect, also in the Norwegian top leadership. The proposal to end financial support to the modernization was dropped before it came to deliberations within the government.[90] Possibly, this was done out of a genuine Norwegian belief that the project was still viable. More likely, however, the decisive factor was apprehension that Norwegian withdrawal of support would lead critics to blame Norway for the collapse of the high-profile Pechenganikel project.

Whatever Norway's reasons for upholding its funding commitment, a contract between Kværner Engineering and Pechenganikel was completed in mid-December and presented to Norilsk Nikel for final approval. In the end, however, Blatov's efforts to keep the modernization plans alive came to nothing. No answer from the Norilsk Nikel concern leadership had been received by the end of 1996,[91] and so, in January 1997, the NME decided against funding Kværner Engineering's further activities in the matter.[92] The first signal from the Norilsk Nikel top management came when Yuri Kotlyar, Norilsk Nikel's second in command, met with the Norwegian ambassador in Moscow. Kotlyar's assessment was that any idea of a modernization at a price of USD 260 million was impossible to entertain, for the crisis-ridden Russian government and the even more depressed Norilsk Nikel concern, which by then had been temporarily placed under state administration.[93] Reportedly, Kotlyar characterized the modernization as 'an over-idealized project that can never be realized'.[94]

The final and formal end to Kværner Engineering's project came in April 1997, in a fax from Aleksandr Khloponin. The managing director of Norilsk Nikel expressed gratitude for the project plans that had been developed, but went on to claim that 'serious change in the conceptual approach to the question of processing ore from Norilsk at the Pechenganikel combine' necessitated new plans.[95] This linguistic trickery could not conceal the basic fact that Norilsk Nikel simply could not be bothered with Norwegian environmental concerns. Further, Khloponin reintroduced the question of Norilsk ore versus local ore. Of vital importance to the technological design of the modernization project, this question had caused difficulties several times, as we have seen. In spring 1994, Pechenganikel had decided to plan for processing of solely local ore. This had changed in November 1995, when up to 300,000 tons of Norilsk ore were projected to be smelted in Nikel annually. Now Khloponin made yet another turnabout and maintained that Pechenganikel would process only local ore. Thus, after years of fatiguing negotiations and high-level diplomacy, the top manager of Norilsk Nikel dismissed all the Norwegian efforts with a brief half-page note.

## Intermezzo: 1997–2000

Khloponin's fax, however, did not mean the end to the modernization saga. While rejecting the outlined project, Khloponin also informed that Norilsk Nikel was in deliberations with the NIB about the viability of a cheaper solution. The new project, he stated, would decrease emissions by 90 per cent.[96] Although Khloponin did not make direct mention of the possibility of Norwegian support, there is much to suggest that he brought up the expected environmental effects of the new project with a view to reactivating Norwegian funding in the future. And, sure enough, Khloponin's intimations were soon transformed into concrete requests. In the summer of 1997, the NME was informed by the NIB about the ongoing negotiations, which in addition to themselves and Pechenganikel involved Norilsk Nikel's engineering branch Gipronikel as well as Boliden Contech, the latter now acting on behalf of the formally non-existent PRC. The new project took as its point of departure one of the alternative modernization plans previously outlined by the consortium. The NIB reported that 'the cooperation of the Norwegian state is very important for both the Norilsk Nikel concern and for Russian authorities'.[97]

Faced with this new and probably unexpected request from a Russian enterprise that on numerous occasions over the years had demonstrated unpredictability and utter disregard of the environmental aspects of its industrial activities, the Norwegian authorities adopted a wait-and-see attitude. While they were obviously chary of embracing the Russian plans unreservedly, there were also several political factors that made outright rejection unwise, as can be seen from this memo:

> The uncertainty regarding the Russian plans and the [Russian] ability to complete a venture of this kind suggests that a possible new project must be comprehensively evaluated by an accountable and independent consultant before [Norwegian] support is contemplated. If we remove the . . . allowance from the national budget, however, we will risk reactions from both [Pechenga's neighbouring municipality] Sør-Varanger, which finds itself in a difficult situation [high unemployment rates due to the demise of the cornerstone enterprise AS Sydvaranger] and from political circles claiming that the Government is downsizing its environmental efforts in Eastern Finnmark and Northwest Russia.[98]

The conclusion, then, was to maintain the promise of support in the national budget. It was clear, however, that persuading the Norwegian government would henceforth be far more difficult than before. Perhaps to the relief of its officials, the NME no longer had to deal directly with the Russian industrialists. As of spring 1997, negotiations were taken care of by the NIB, with regular updates to the ministry. In effect, the role of the Norwegian authorities was limited to be a potential funding source, and thus a far less active participant in various project development schemes.

Fairly soon, it became clear that the waters were muddy also this time around. The Norilsk Nikel concern, which became fully privatized in August 1997,[99] continued to outline alternative approaches to the modernization scheme. This must be understood against the backdrop of recent changes in Norilsk Nikel's boardroom. The new private owners faced many strategic choices for the development of the concern, and Pechenganikel had to be part of that bigger picture.[100] Any modernization there would have to be accommodated to other strategic decisions made for the Norilsk Nikel concern as a whole.

These were turbulent times for Pechenganikel, and the future was uncertain. To attract investment capital, the Russian nickel industry searched far and wide for business opportunities. From fall 1997 to summer 1998, Norilsk Nikel was in negotiations with several Western companies for possible collaborative efforts in Pechenga.[101] Pechenganikel's main problem was, as phrased in a local newspaper, that 'we [Pechenganikel] have metal, but are short of money'.[102] In other words, there were simply not enough investment funds available for further development the remaining ore deposits. The closest Pechenganikel came to realize any major collaboration in this period was an agreement of intention with Finnish Outokumpu, whereby the two companies set up a joint venture to develop mining operations in Pechenga. Outokumpu was still very interested in access to Russian raw material to feed its stainless steel production (see Chapter 3). Pechenganikel, on the other side, needed any help it could get in extracting more ore for its smelters.[103]

For Pechenganikel, a company short on cash but still reportedly rich in resources,[104] the main goal in the late 1990s was to secure further production. Considerable investments were necessary to extract more ore, and the increasingly run-down factory complex was in dire need of upgrading. In this picture, Norwegian support, which was intended to facilitate discharge cuts, was also seen by the Pechenganikel management as a potential funding source for necessary maintenance and upgrading at the smelters. There was little or nothing to indicate that the new Pechenganikel management actually took an interest in the environmental aspects of a possible modernization. If, however, a modernization effort could raise the standards at the smelter and thereby enable future production, Norwegian and Russian interests converged. In talks with the NIB, Pechenganikel repeatedly emphasized that reconstruction of the smelter in Nikel would depend on Norwegian financial support.[105]

However, events were to render both Pechenganikel's mining collaboration with Outokumpu and the modernization project temporarily impracticable. August 1998 was a watershed month, both for the modernization as such and for Pechenganikel as an industrial enterprise. In mid-August the proposed modernization project was assessed and found lacking by the Norwegian authorities. According to the Norwegian Pollution Control Authority (SFT), the projected reduction in emissions presupposed an 'idealized situation in which there is no room for disturbances of production'. SFT, apparently, had no reason to believe that such an ideal state existed at Pechenganikel and consequently dismissed the proposal.[106]

Far more important than the negative assessment conclusion from Norwegian authorities, though, was the massive financial crisis that had been looming for a while before striking the Russian economy with full force in late summer 1998. When the Russian government on 17 August decided to devalue the rouble, it was an act that would have far-reaching consequences, also for Pechenganikel. The ensuing financial crash, although the relative importance of its endogenous causes (insufficient Russian fiscal discipline) and exogenous causes (the crisis in the Asian markets) was heatedly debated by economic experts,[107] laid bare the many structural deficiencies of the Russian economy. The burden of unsustainable loan-financed public expenditure and especially over-spending in the regions, which had been given wide-reaching fiscal and political freedoms in the Yeltsin period, became too much to bear for the volatile Russian currency. Thus, the rouble was devalued as a last resort after the State Duma voted against government proposals to redirect tax revenues from the regions to central coffers.[108]

There was, as this episode illustrated, little that the central level could do to bring the regional leaders into its fold.

For the Russian nickel industry, the effects of the deteriorating economy had been palpable already before the rouble was devalued in August. While one can argue that the crisis was Russian in its essence,[109] there was still no denying that the breakdown in Asian markets and the ensuing fall in demand for raw materials in general had a negative effect on the nickel-exporting companies of the Kola Peninsula.[110] In November 1998, the Norwegian consulate in Murmansk reported that, especially at Severonikel in Monchegorsk, the economic downturn was sorely felt, as workers were not paid their wages – or were laid off. Although Pechenganikel had still avoided the worst of the crisis, the same structural problems applied to Nikel and Zapolyarnyi. The Norilsk Nikel management had not been sitting idle, though. On the top level, Oneksimbank, which had controlled Norilsk Nikel shares for about a year, was in talks with two other Moscow-based banks, Menatep and Mostbank, with a view to merger with them. The holdings on the Kola Peninsula underwent restructuring as well. Severonikel, Pechenganikel and a supporting mechanical plant in Olenegorsk, just north of Monchegorsk, were consolidated into one unit. The establishment of the Kola mining and metallurgical company (KGMK) was presented to the public on 3 November 1998 by the Norilsk Nikel director Khloponin and the governor of Murmansk oblast, Yuri Evdokimov.[111]

The governor's presence reflected a stronger regional commitment to support the local nickel industry, and thus marked a shift in the relationship between the privatized companies and the authorities. As noted above, companies and public authorities had, since the denationalization process started in the mid-1990s, been bickering about taxes and social responsibilities in the company towns. Due to the crisis situation that arose in Murmansk oblast cornerstone companies in 1998, major social upheaval threatened. Both Pechenganikel and Severonikel were still funding all schools, nurseries, healthcare and other public services in their local communities, and the survival of the companies was essential for the continued existence of Monchegorsk, Zapolyarnyi and Nikel.[112] The oblast administration had no option but to agree to give the new company substantial tax breaks. The oblast also assumed full responsibility for public services in the towns after a period of two years. Finally, the oblast administration bought 10 per cent of the shares in KGMK – an investment that had become even more important since Outokumpu had withdrawn from further collaboration with Pechenganikel after the Russian rouble crisis had set in.[113]

Despite the immense difficulties in the nickel industry, representatives from Norilsk Nikel and Pechenganikel continued to hold talks with the NIB, and the latter reported to the Norwegians that the crisis had not dampened the concern's willingness to carry on with the modernization plans.[114] Progress was slow, however, and no new movement was seen in the project until the beginning of 2000, when a markedly improved situation in the world nickel market was conducive to investment in the industry. Pechenganikel was getting back on its feet, and signalled preparedness to go ahead with a modified project with a price tag of USD 60 million, based on technology delivered by Norwegian Elkem and Swedish Boliden.[115] This scheme proved short-lived, however, as a new turnaround was imminent. In spring 2000, Norilsk Nikel declared that a modernization would have to be carried out based on 'Soviet' technology, which according to it would meet both financial and environmental demands in a better way than the Swedish–Norwegian solution.[116] This is a development we will examine further in Chapter 5.

## The 1990s: A state of emergency

As we have seen, the Pechenganikel modernization project went through several phases in the course of the 1990s. At the outset, it was surrounded by strong optimism, and seen as a demonstration of a new age in Russian–Nordic relations after the Cold War. Not only would it show that the Nordic states could collaborate effectively with post-Soviet Russia in a massive undertaking in one of Russia's major industrial branches – it would address a shared environmental concern and thereby serve as an example of a new mutual understanding of the need to protect the fragile Arctic environment. Disappointment loomed, however. Post-Iron Curtain optimism soon gave way to a realization that the deep Nordic–Soviet chasm in approaches to the natural environment and to business relations had not disappeared with the 'liberation of Russia'. Pechenganikel modernization soon became a long-drawn and dreary bilateral matter between Norway and Russia rather than a multilateral flagship project between the 'northerners of Europe'.[117] It is within the framework of Nordic, especially Norwegian, strategies on how to relate to a new and unpredictable Russia that the Pechenganikel modernization must be understood.

As mentioned, the 1990s were a time when the small state of Norway became a donor vis-à-vis its giant neighbour Russia. This historic anomaly in bilateral Russian–Norwegian relations was made possible by the sudden collapse of the Soviet Union and the ensuing socio-economic crisis in post-Soviet Russia. The many unprecedented processes that took place in Russian domestic and foreign politics during this time hinted at a wide-ranging disintegration of both formal and informal norms. With the collapse of the Soviet Union, Russian society entered a ten-year 'state of emergency' in which extraordinary developments could occur. In the following, we will look at some of the processes that took place and that to a lesser or greater extent had a bearing on the Pechenganikel modernization – in itself a striking example of how open the bilateral Russian–Norwegian playing field had become after the collapse of the Soviet Union.

One overarching element in the Norwegian political response to the Soviet collapse and the ensuing opening up of post-Soviet Russia was region building. BEAR, also referred to as the Barents cooperation, was launched in January 1993. BEAR involved northern regions in Russia, Finland, Sweden and Norway, while its cooperative bodies also included representation from Denmark, Iceland and the European Commission.[118] The cooperation was the result of a Norwegian initiative and was infused with the euphoric rhetoric of the early post-Soviet years. The intention was to create a platform for the anticipated explosion of East–West contact after the lifting of the Iron Curtain. Two prime target areas of BEAR were environmental protection and expansion of business relations in the Barents region[119] – both central elements in the Pechenganikel modernization scheme.

Unsurprisingly, then, the Norwegian authorities saw an opportunity to link the two. Soon after the first public mention of establishing a regional cooperative body in the North was made in April 1992,[120] the Ministry of the Environment considered the Pechenganikel modernization as 'a first big step in the development of the regional Barents cooperation'.[121] At this time, the first modernization project developed by Finnish Outokumpu was collapsing, due to the high price tag, but that only seems to have spurred Norwegian authorities. Finnish failure opened up for Norwegian industrialists to take Outokumpu's place, thereby securing the Norwegian authorities an even more influential position in the modernization of the nickel complex and, consequently, in the region-building efforts. This was essential because the project itself was expected to help Norway achieve what all Norwegian political camps at the time saw as an important foreign policy goal. Buoyed by the widespread Barents euphoria, the NME hoped that the modernization could prove 'one of the cornerstones in the development of the cooperation in the Barents region',[122] playing a pivotal role in their ongoing region-building efforts.

Although the Pechenganikel modernization always was an important undertaking in its own right, it is likely that Norwegian interest was boosted by ambitions of strengthening the BEAR cooperation. Several references to how the Pechenganikel project was to provide BEAR with concrete content in the form of a commercially and environmentally beneficial undertaking were repeatedly made in the subsequent years.[123] However, as the Barents euphoria of the early 1990s gave way to a far more critical discourse in the latter half of the decade, the Pechenganikel modernization became just another example of Russian–Norwegian projects gone sour. By the end of the decade, BEAR had arguably become a largely Norwegian-run venture with only nominal support from the other participating countries, and not the dynamic multilateral meeting place it was intended to be.[124] Moreover, linkages between the modernization project and the Barents cooperation were rarely mentioned in the Norwegian bureaucracy. Pechenganikel had failed to emerge as a shining example of Russian–Norwegian cooperative spirit, and the Barents cooperation had scant relevance for further efforts to curtail sulphur emissions in the Pasvik valley.

Instead, an increasing number of critical voices were raised in Norway against further funding of modernization projects at Pechenganikel. The main argument was that the Russian nickel industry and Norilsk Nikel in particular, was becoming highly profitable in the hands of private owners and should therefore be expected to finance their own clean-up operations. This criticism was connected to the emergence of a new phenomenon in Russia: the growing number of individuals who had built their fortunes in the years of the 'state of emergency' through acquiring state-owned businesses. What is widely seen as the uncontrolled and unlawful privatization of Russian enterprises and the ensuing political influence of these 'oligarchs' certainly affected Norilsk Nikel and, therefore, also Pechenganikel.[125]

In the case of Norilsk Nikel, it was Vladimir Potanin, a financial trader born and bred within the Soviet elite, who after protracted manoeuvrings took over ownership of the previously state-owned enterprise. As the owner of Oneksimbank, Potanin acquired control over the Norilsk Nikel concern through a highly controversial shares-for-credit arrangement with the Russian state, entered into in 1995 and finalized in 1997.[126] Potanin even held high political office for a while.[127] Thus, he was a typical representative of the new order of post-Soviet and well-connected businessmen, who not only had quickly amassed immense wealth but also were occupying political positions. As a result, Russian industrial and financial activities from the mid-1990s were governed less by law than by the personal ambitions of a limited number of powerful individuals who were able to navigate and influence the still volatile post-Soviet economy.

To what extent this development affected the Pechenganikel modernization as such is more difficult to ascertain. It might be tempting to claim that the oligarchical takeover of Norilsk Nikel represented a step towards the breakneck pursuit of profit with little or no regard for the environmental effects of production.[128] As probable as this may seem, there is little to indicate that continued state ownership would have proven more ecologically sound. As we have seen, early attempts at achieving emissions reductions while the Pechenganikel enterprise was still under Soviet and later Russian state control brought little success. This can be ascribed to the instability in the Russian political sphere for much of the 1990s, as well as the many transitional problems associated with market reforms. However, as shown in this chapter, Russian authorities only haltingly and halfheartedly if at all – and only when repeatedly called upon – provided the necessary framework for successful completion of the Pechenganikel modernization. Conversely, one can even argue that, as regards environmental concerns, the Norilsk Nikel privatization and Vladimir Potanin's takeover represented a change for the better.[129] Replacing an inert management that was still floundering in the quagmire of Soviet traditions,[130] the new private ownership at the very least represented a fresh start. As we shall see in Chapter 5, the rejuvenation of Norilsk Nikel eventually did bring about a Russian initiative to modernize the installations at Pechenganikel.

It seems unwise, then, to conclude that privatization necessarily meant that Norilsk Nikel became less environmentally friendly than it would have been under state ownership. It is equally imprudent to conclude that oligarchical control over Norilsk Nikel resulted in increased instability in the concern, obstructing the Pechenganikel modernization. In fact, one can argue that the privatization brought more rather than less predictability to the Pechenganikel management. In terms of business organization, Potanin's takeover was certainly beneficial to Norilsk Nikel and helped transform the concern from post-Soviet wreckage to the highly profitable multinational giant that it is today.[131]

That said, turnover and instability in the Russian company would continue to hamper progress in the Pechenganikel modernization also after privatization, as will be shown in Chapter 5. The question is whether this was a result of the privatization as such, or a reflection of the general turmoil in Russian business and society during the long transition period. Arguably, instability within Norilsk Nikel would have been equally or more pronounced if the Russian state still held ownership, as purely political considerations would then have cast even larger shadows over corporate decision making. Under Potanin, there was at least a clearly stated goal for the nickel industry on the Kola Peninsula: to make as much money as quickly as possible – a completely acceptable aim for any privately owned industrial enterprise. Any environmental efforts would be the result either of irresistible outside pressure or, indirectly, of upgrades made in order to improve efficiency. As we shall see in the next chapter, the Norwegians eventually came to realize that they would not be able to apply the necessary pressure, and that there was little help to be had from Russian authorities. And so, they chose the second option – to let the Norilsk Nikel concern take the reins in the modernization project, and develop a technological scheme based in Pechenganikel's industrial needs for an upgrade. With the help of their still-valid promise for funding, the Norwegians hoped to ensure that the refurbishments would lead to an environmentally acceptable solution.

\* \* \*

A state of emergency is defined not only by its extraordinary state of affairs, but also by its impermanence. Sooner rather than later, it must come to an end, and be followed by a new normalcy. This was also the case in Russia, whose post-Soviet 'state of emergency' carried within it the seeds of the new normalcy from the outset. Already in 1992, the historian and Russian studies expert Stephen F. Cohen coined the phrase 'the cold peace' in a scholarly article.[132] The term refers to the conviction in some quarters of Russian society that the termination of the Cold War did not entail an end to Western ambitions of weakening

Russian statehood. Cohen observed that more and more Russians were growing increasingly disillusioned with Russian-style democracy and the failed shock therapy for reforming the economy. Democratization and market reform came to be seen as harmful imports from the West, as measures in fact designed to halt Russian progress rather than to bring prosperity.

Also, in bilateral Russian–Norwegian relations, sentiments to this effect became increasingly pronounced as the 1990s progressed.[133] Russian recipients of Norwegian aid, and perhaps most often Russians who were not on the receiving end, expressed a growing concern with what they perceived as a hidden agenda behind Norwegian project funding. Although this did not come directly into play in the case of the Pechenganikel modernization, it certainly coloured the general Russian–Norwegian cooperation climate towards the end of the 1990s and further into the next decade. The Russians wanted to be in the driver's seat on the road to their own future. With the turn of the millennium, both the political and economic framework in Russia allowed for a more assertive stance in international affairs. As we shall see in next chapter, this was to have a bearing on Russia's overarching foreign and domestic policies – but it also deeply and directly affected the Pechenganikel modernization project.

# A Russian revival

For Russia, leaving the last decade of the twentieth century also meant emerging from the 'state of emergency'. The painful 1990s, when the Soviet mastodon had been reduced to a debt-ridden 'sick man' of Europe, were finally over. Not only was the economy bouncing back, but a new leader who enjoyed immense popularity after first overseeing a brutally successful military campaign in Chechnya as prime minister and then commencing an overhaul of the Russian public sector as acting president,[1] was personifying and legitimizing this Russian revival. With Boris Yeltsin's retirement on New Year's Eve 1999 and the simultaneous appointment of the new acting president, a new era had begun – the era of Vladimir Putin. Putin proved, unexpectedly to most observers, able to assert himself quickly in the face of various domestic pressure groups, including the oligarchs. He would soon develop into the single most dominant political force of Russia.

As the influential new leader that he turned out to be, Putin was able not only to direct Russia out of the 'state of emergency', but also to re-establish a normalcy in the country's relationship to other states. The new normalcy was shaped by the tumultuous 1990s in many ways, and it would be unwise to conceptualize it as a reversal to past practices. Putin's ascendancy implied neither a full-fledged return to totalitarian power principles at home, nor a complete revival of Soviet-style confrontation with the West. Post-Soviet Russia had, after all, been irreversibly changed by ten years of openly interacting with the outside world and was as touched by globalization as any other country. Nevertheless, a certain set of traditional Russian leadership principles, reflections from both Bolshevik and pre-revolutionary practices, were re-invented in Putin's mode of governance. Domestically, he firmly reinstated the Kremlin as a dominant central power by bolstering the vertical command structure so characteristic of Russian governance since the reign of Ivan IV from the mid-1500s. On the international stage, Putin's policies aimed, through a series of pragmatic alliances, at restoring Russia's position among the world's leading powers. Aided by soaring world market prices for petroleum and other raw materials, including nickel, he was able to complete this political project swiftly.[2]

Naturally, the bilateral Russian–Norwegian relationship, which had formed the primary political context for the Pechenganikel modernization project, was affected by the thriving development Russia enjoyed during Putin's presidency. Also here, a new normalcy soon emerged. Unlike what the Norwegian authorities had become accustomed to during the 'state of emergency', Russia was no longer an impoverished neighbour in despair, but rather the daunting great power of the past. As in any other relationship between a great power and a small state, the stronger party emphatically called the shots, as was the historical norm in Russian–Norwegian relations. Although awareness of the new reality, or rather the re-establishment of the historical imbalance between the two states, dawned on the Norwegian bureaucracy only gradually, the effects of Russian revitalization on the Pechenganikel project became evident fairly soon after the turn of the millennium. This is not to say that the Russian state intervened in the project, nor that affluent Norway was rejected outright. Norwegian money, it seems, was still very much coveted by Norilsk Nikel. From now on, however, it was the concern itself that would aim to control the process.

## Turning the tables

As noted in the previous chapter, the Norilsk Nikel concern declared in spring 2000 that the modernization of Pechenganikel should be based on its own technology. The concern's insistence that Russian know-how would provide the best solution to Pechenganikel's emission problems reflected more than merely a technological preference. It can arguably be understood as one of many expressions of a new-found Russian self-esteem that became apparent from the turn of the millennium. With Vladimir Putin's resurrection of Russia's international standing permeating the country's self-understanding, the time had come for Russia to shed the passive role of aid recipient and assume control over its internal affairs. Doubtless, this shift in the Russian approach to the outside world also affected the Norilsk Nikel leadership and can help explain the more aggressive line now taken by the firm in its dealings with the Norwegians. That said, Norilsk Nikel still felt the lure of Norwegian funding and was not ready to completely dismiss the gift package.

But Norilsk Nikel's offensive was based on more than just an abstract rebuilding of national pride. The company was thriving, thanks to rising nickel prices on the world market, and probably also to organizational rethinking provoked by the financial breakdown in 1998. Also, to Pechenganikel as to any other exporting business in Russia, the August 1998 devaluation was highly advantageous in the longer term, as export incomes received in foreign currency became substantially higher when exchanged into the devalued domestic rouble. A remarkable reflection of these factors was that the price of Norilsk Nikel shares skyrocketed from USD 0.70 in 1999 to USD 11 in 2000. The positive development also had effects on the Kola Peninsula. From being doomed to decline and likely bankruptcy within a decade, as was the verdict in mid-1999, the prospects of both KGMK and Pechenganikel were considered to be very bright a mere twelve months later. In Zapolyarnyi there was talk of comprehensive investments in the mining operation that might prolong Pechenganikel's lifetime until at least 2030.[3]

As it turned out, the new technological solution advanced by Norilsk Nikel was not novel at all. Once again, it was the technology developed by Vladimir Vanyukov that was promoted by the concern. When the modernization of Pechenganikel first came up in the late 1980s and early 1990s, Soviet representatives had repeatedly proposed installing Vanyukov furnaces. These furnaces, which had been fitted in Norilsk

Interior of the nickel smelter. The proposed Vanyukov technology aimed at curbing all sulphuric emissions from the smelting process.
Photo: Ola Solvang

copper smelting operations in the mid-1980s, were then dismissed as insufficiently environmental-friendly by Norwegian companies and authorities alike (see Chapter 3).

There is no trace of any technical assessment to support this position, and Norway's dismissal of the Vanyukov technology requires a different explanation. The motivation of the Norwegian companies seems obvious: fending off competition and landing a lucrative long-term contract in the then-emerging Russian market. Similarly, the Norwegian authorities were set on channelling as much as possible of the Norwegian funding back to domestic businesses. The rejection of the Vanyukov technology, then, was probably not based on environmental or technological reasons alone but was also motivated by business considerations. To the public, the rejection was explained without ambivalence: Only Western (preferably Norwegian) technology could provide an 'environmentally satisfactory solution'.[4]

Considering the earlier dismissal of the Vanyukov furnace as being underdeveloped and un-environmental, Nordic assessments in 2000 make for interesting reading. These documents strengthen the assumption that the Vanyukov technology had been rejected in the early days of the modernization saga not because it failed to measure up technologically or environmentally. Boliden Contech, the Swedish company that had been part of PRC in the 1990s and that was still involved in the modernization scheme, was asked to examine the sketches. The Swedes quickly concluded that the Russian solution was better than their own proposal. Similarly, a consultant from the Nordic Environment Finance Corporation (NEFCO) who had been involved in the modernization efforts throughout the 1990s stated that the Vanyukov technology was environmentally superior to all other solutions.[5] These estimates alone were not enough to ensure Norwegian funding, but they certainly made it difficult to reject the Russian Vanyukov technology once again.

At the time, the Norwegian Parliament was debating whether to shelve the whole modernization project. The opposition parties doubted its realism, due to the many delays, and opened up for reallocation of the 270 million NOK that remained after the first decade of project development and failed negotiations.[6] That meant that the NME was put under pressure: it would have either to reject further attempts at realizing the modernization, or to argue the importance of continued efforts.[7]

More remarkable, however, and symptomatic of the newly-won Russian self-confidence, was the pressure that came from Norilsk Nikel. In a press release in early August 2000, the concern announced that a development plan for Pechenganikel up to the year 2015 had been approved by its board of directors. Emphasizing the environmental improvements that would result from this upgrading, the press release stated that the planned modernization of the smelter and sulphuric acid plant in Pechenga, priced at just under USD 70 million, would probably be funded by a grant from the Norwegian government and loans from NIB.[8] Interestingly, this very public announcement of

Norwegian willingness to fund the project was made without any prior Norwegian approval and long before Norwegian authorities had had a chance to assess the Vanyukov technology properly.[9] Norilsk Nikel had taken the initiative: instead of being subjected to Norwegian pressure to modernize and cut emissions, the concern now presented a firm plan to this end, while expressly expecting the Norwegian authorities to make good on their earlier funding pledge. The tables had indeed been turned.

In Oslo, the Ministry of the Environment, finding itself under pressure from several directions, was compelled to make a decision. Should they abandon the Pechenganikel modernization completely? The ministry asked the SFT to weigh further attempts to modernize Pechenganikel against redirecting the funds for a similar modernization at Severonikel in Monchegorsk. Since reconstruction at Severonikel would be likely to encounter the same problems as those experienced in Nikel, and considering that emissions from Monchegorsk did not affect Norwegian territory as much as did the Pechenganikel emissions, the SFT's conclusion was unsurprisingly firm: only further efforts to modernize Pechenganikel would make any sense, especially taking into account the promising Russian proposal to implement Vanyukov furnaces. On the other hand, the SFT made it clear that any Norwegian support to the Russian proposal should be granted only on condition that the Vanyukov technology was thoroughly scrutinized and that such support be subjected to a range of contractual provisions that would ensure environmental improvements in and around Nikel.[10]

The four interested parties – Norilsk Nikel/Pechenganikel, Boliden Contech, NIB and the NME – agreed to establish an assessment group to examine the feasibility of installing Vanyukov furnaces in the Pechenganikel smelter and whether this solution would provide the required reductions in emissions. Norilsk Nikel, mindful of the sensitive nature of its technology, wanted to contain information about the Vanyukov furnace as much as possible.[11] Therefore, the PERG consisted of only four members, each stakeholder appointing one, all of whom were required to abide by a series of confidentiality regulations.[12]

Based on documentation provided by Norilsk Nikel, and after visits to Norilsk, Moscow and Pechenga, PERG approved the modernization plans. However, the group called for extensive control mechanisms, both during project implementation and after project completion, to monitor anticipated improvements of the environmental situation in the area.[13] Thus, the stage was set for a final international effort to better the environment of the Pasvik valley.

At the turn of the millennium, as we have seen, the situation was quite different from what it had been throughout the previous decade and even a few years earlier. Not only had Pechenganikel started making money, but the Russian industrialists had managed to promote their own technology as the cheapest and most efficient remedy to the environmental disruption. Norwegian government agencies like the Ministry of the Environment and SFT were, despite the many disappointments of the past decade and the mounting scepticism within the parliament, still intent on upholding their promised support to the modernization effort. However, recent experience had prompted the Norwegians to hedge their bets. As we shall see in the following, a whole cluster of agreements would be set up to safeguard the Norwegian contribution to Pechenganikel.

## Suspicious minds

A new modernization effort, based on a different technology and with some different actors involved, called for new contract negotiations. The NME, having experienced many unexpected twists and turns in its dealings with Russian authorities and industrial actors in the Pechenganikel saga, was now intent on setting up a contractual framework that would protect its interests. It seems, also, that the ministry preferred a position as a mere donor rather than a controller: the Norwegians wanted to maintain some distance to the new project. Even before PERG's approval was declared in January 2001, the NME had approached NIB with the idea of a trust fund. The plan was to transfer the Norwegian contribution of 270 million NOK to the bank,[14] leaving NIB with project administration responsibility and relieving Norwegian authorities of direct contact with Norilsk Nikel/Pechenganikel.[15] Instead of being a matter handled jointly by Norwegian and Russian authorities, from now on, at least in terms of daily management, the modernization would become a matter between NIB and Norilsk Nikel/Pechenganikel, with Boliden Contech as a side partner. NIB was of course expected to report regularly on progress to the NME, while Norilsk Nikel/Pechenganikel would communicate directly only with NIB.[16]

The sources do not explicitly show why the Norwegians adopted this new approach. However, it seems reasonable to assume that a certain level of project fatigue within the Ministry of the Environment played a role. One decade of optimistic peaks repeatedly undermined by unexpected changes on the Russian side must have taken its toll. More important, though, were the legal aspects. The Norwegian authorities obviously wanted to cloak the new project in legal provisions that would provide them with an exit if that should prove necessary. After all, Norwegian experiences from the recent past gave scant grounds for optimism with regard to Russian predictability and reliability. NIB would provide Norwegian authorities with a buffer zone, so to speak, and the greater distance to the project itself would lessen the chances of Norway having to make concessions towards the Russians. Furthermore, by taking the project out of the Russian–Norwegian bilateral realm and placing it within a strictly commercial and industrial context, one could hope that the effort would be shielded from changing political winds on either side of the border. The non-profit aspect, represented by environmental concerns, was relegated to the background and overtaken by a legally stringent relationship between a bank, NIB, and an industrial actor, Norilsk Nikel/Pechenganikel. Although Norwegian conditions for releasing funding – a 90 per cent reduction of emission levels at Pechenganikel – stood firm, the project could be nominally governed by profitability and industrial needs rather than the original motivation of bettering the environment.

Throughout 2001, various Norwegian government agencies thoroughly assessed, discussed and evaluated the legal arrangements surrounding the new modernization project.[17] Norwegian government agencies were party to the agreements that were finalized first. On 19 June 2001, the Russian government agreed to exempt Norwegian funding of the modernization project from federal taxes.[18] Along the same lines the governor of Murmansk oblast, Yuri Yevdokimov, had guaranteed four days earlier that no regional taxes would apply to the funding provided by Norwegian authorities.[19] Despite these written promises, the Norwegians probed even further among Russian decision makers to make absolutely sure that the assistance would not be depleted by any form of Russian taxes. In an interview with the chairman of the Murmansk Regional Duma in September, the Norwegian Consul-General received additional guarantees of tax exemptions, both regional as well as local levels.[20] This Norwegian caution and suspicion, and the scepticism with which the various assurances were received, demonstrates the deep-rooted wariness that had developed in Norway over the years. Clearly, Russian–Norwegian relations could not be described as particularly full of trust at this time. Moreover, Norwegian government circles had become acutely aware of the multitude of Russian players with varying interests, governmental and industrial alike, with the potential to halt progress or even threaten the project.[21]

The same caginess characterized the Norwegian approach to the other agreements required to get the modernization underway. Although the Norwegian authorities would be a party to only one of them, between themselves and NIB, they carefully scrutinized NIB's two agreements with Norilsk Nikel/Pechenganikel.[22] One of these, the Grant Facility Agreement, regulated Pechenganikel's access to the Norwegian funds managed by NIB, whose counterpart in this agreement was KGMK, the Kola mining and metallurgical company.[23] The other, the Support, Share Retention and Guarantee Agreement, was between Norilsk Nikel as mother company to KGMK (and Pechenganikel), and NIB. According to this agreement, Norilsk Nikel promised to keep all KGMK shares on their hands and to guarantee KGMK's obligations described in the

Grant Facility Agreement, as well as KGMK's responsibilities towards NIB described in their loan agreement.[24] In addition, a technical annex was signed by Norilsk Nikel and NIB. A comprehensive document of eighty-six pages, it covered all aspects of the modernization project – from technological solutions to loan terms and environmental requirements.[25]

When signed in late December 2001, these contracts provided a legal framework for the upcoming modernization project that was superior to the far less detailed agreements that had existed between Norway and Russia during the turbulent 1990s. Not only were the contracts much better developed, but also they included a Russian party that seemed more accountable than what had been the case during the previous decade. If not firmly stabilized, Norilsk Nikel and KGMK were at least firmly defined as companies with a private owner structure, rather than in limbo between the Soviet state and the new Russian capitalist class. Responsibilities and obligations were well-defined – and, significantly, the duration of the agreements was limited. In both the Russian–Norwegian tax exemption agreement and in the management agreement between the NME and NIB, it was clearly stated that the content would be valid for a maximum of ten years.[26] Thus, although many surprises would still riddle the modernization efforts, there was now a certain degree of legal protection against a protracted process with no result.

The project officially commenced in 2002. However, it rapidly became obvious that the wide-ranging contractual framework described earlier in this chapter was still not enough to put Norwegian worries to rest. Fundamental distrust still seemed to exist, especially at the political level in Norway. In September, in connection with a visit to the Kola Peninsula and Pechenganikel, the Norwegian minister of the environment, Børge Brende, gave an interview to the Norwegian daily *Aftenposten*. Here, he criticized the lack of progress in the modernization efforts and even reflected on whether environmental subsidies were the right approach for achieving improvements in Russian environmental standards.[27] At this time, Norilsk Nikel could boast substantial profits from its export sales and was at the forefront among the world's nickel and palladium producers.[28] This made it pertinent to question the need for Norwegian subsidies to the Russian nickel industry, and Brende's statements must be seen in light of this.

The issue really blew up, however, when the Brende interview was picked up by the Russian newspaper *Novye Izvestiya*. According to the general director of KGMK, Evgeny Romanov, the article in *Novye Izvestiya* more than hinted that Brende had said that Norwegian support funds were being embezzled by the Russians while the environmental disruption kept increasing. Romanov, clearly worried about his company's image, demanded that Brende and his ministry explain its official stance on the matter.[29]

While the NME's denial of *Novye Izvestiya's* own interpretation of Brende's statements was readily accepted by the Russians,[30] there was still the problem of the actual content of the minister's interview with *Aftenposten*. Romanov sent a letter directly to Brende, asking him to explain himself.[31] In the Norilsk Nikel leadership the alarm bells sounded even more strongly. In a fax to NIB and Boliden Contech, Norilsk Nikel's deputy general director A.N. Burukhin referred to 'the blatant distortion of reality' presented by *Aftenposten*, and went on to explain how the recent press coverage had 'caused a negative reaction from Russia's state officials and led to a need for . . . Norilsk Nikel's management to submit appropriate clarifications and refutations of the facts contained therein'. Burukhin's conclusion was dramatic:

> Norilsk Nikel's management is extremely surprised by the negative reaction from Norway's Ministry of Environment Protection [*sic*] upon the visit to our operations. The situation that has emerged is forcing us to take a temporary pause in the implementation of the agreements reached in order to be able to analyze it.[32]

In Oslo, a decision to delay the modernization seemed *too* dramatic. Some officials in the Ministry of the Environment suggested the possibility of a hidden agenda behind Burukhin's statement. Rumour had it that Norilsk Nikel was planning to terminate the activity in

Nikel, with a wholesale transfer of all smelting activities to Severonikel in Monchegorsk – a decision that would in effect give grounds for termination of the Norwegian support to Norilsk Nikel.[33] Thus, Burukhin's reaction led some Norwegian ministry officials to believe that Norilsk Nikel, by raising a ruckus, were buying themselves time to decide on the future of Pechenganikel.[34] These rumours of Pechenganikel's imminent demise, although unsubstantiated, were rekindled and re-appeared regularly. Eventually, the Norwegians decided not to put faith in them.[35]

Whatever the motives behind this commotion, the whole affair certainly demonstrated that the relationship between the Norwegian donor and the Russian industrialists was fraught with communication problems and scepticism. The chasm between Norwegian environmentalist objectives and Russian industrialist motivation was clearer than ever. Before the project had even properly begun, it was undermined by unwise statements by Brende and a petty squabble resulting from an inconsequential newspaper article. The letter that the Norwegian minister of the environment finally sent to Norilsk Nikel's president Mikhail Prokhorov could hardly be considered an apology or an explanation of his statements. Rather, after having summarily reiterated his continued support to the project, Brende let it politely be known that he expected Prokhorov to agree with him on the virtue of re-investing industrial profit in environmental improvements and factory modernizations that would enhance the working environment for the staff. He also mentioned that he was aware of Norilsk Nikel's immense earnings over the last few years.[36]

Already at the outset, the project had been delayed for six months. Contract negotiations between NIB and Norilsk Nikel took longer than expected, and the quarrel described earlier must have slowed down the project during its initial phase in the fall of 2002. Soon it also became clear to NIB that there was still a way to go before Norilsk Nikel and KGMK/Pechenganikel could be considered stable counterparts. In every project report from NIB during the first few years of the project, the numerous delays are explained, inter alia, by the 'complex organizational structure' in the Russian industrial concern and the internal communication difficulties ensuing from this. Turnover of key personnel is repeatedly mentioned as a problem.[37] Thus, in the first report from the beginning of 2003, NIB noted that the project was developing satisfactorily, 'though not as planned'.[38] In the coming years, the modernization venture would never quite recover from its unfortunate start, and before long the whole endeavour would be jeopardized by repeated postponements.

## Delays

At this point it is relevant to examine the main technical aspects of the modernization project. Basically, this was a three-pronged effort focused on three different stages in the ore refining process. Firstly, the initial treatment of the ore, the pellet-roasting, which was undertaken in Zapolyarnyi, was to be replaced by cold briquetting technology. This meant that the ore was to be compressed rather than roasted, and instead of substantial sulphur dioxide emissions – the result of the traditional roasting method – most of the sulphur naturally contained in the ore would not be released at this stage. This improvement would reduce sulphur dioxide emissions in Zapolyarnyi to a minimum. However, the sulphur would not vanish but still be enclosed in the briquettes when they reached Nikel for further refinement. Therefore, in Nikel, the original electrical smelters and convertors – which had not been designed to handle sulphur-rich ore in an environmentally sound way – would be gradually decommissioned and replaced by two-zoned Vanyukov furnaces. These Vanyukov furnaces were designed to complete the refining of the ore into nickel matte (with several bi-products, such as copper and palladium) in a physically closed process with a minimum of emissions. At this stage of the refining process all the sulphur in the ore would be released but confined within the furnace structures. This made the third and last main component of the modernization effort, the new sulphuric acid plant, crucial indeed. This plant had to be capable of handling the immense amounts of sulphur that were no longer to be emitted in Zapolyarnyi, through Pechenganikel's smokestacks or peripherally throughout the refinement process, but that were now captured in the Vanyukov furnaces. In the sulphuric acid plant, this captured sulphur would be liquefied. The projected result of the modernization, then, would be a 90 per cent reduction in aerial emissions of sulphur dioxide (measured from the 1999 emission levels) and a correspondingly large increase in Pechenganikel's production of sulphuric acid.[39]

It was problems related to offloading the large amounts of sulphuric acid that created the first major delay in the project. During the planning period, a central point had been to find a reliable market for the increased production of sulphuric acid at Pechenganikel. Conveniently, a large apatite enterprise located in Apatity further south on the Kola Peninsula had pending plans to establish a fertilizer production facility. For this production, the enterprise would need large amounts of sulphuric acid, and therefore presented itself as a suitable buyer of Pechenganikel's side-product. By the summer 2003, however, the Apatity combine had abandoned their plans. Pechenganikel now felt compelled, or so the matter was presented, to re-assess the modernization plans. There were doubts as to whether the Vanyukov furnace, with its massive generation of sulphuric acid, would be suitable in a situation with no immediate apparent market for the acid.[40]

During the next six months – late summer, fall and early winter 2003/04 – the possibility of abandoning the Vanyukov solution and instead upgrading Pechenganikel's existing convertors and furnaces was discussed among the Russian stakeholders. It was argued that both alternatives would entail delays, but that securing new buyers for the sulphuric acid produced in the Vanyukov furnaces would be more time-consuming. By August 2003, it seemed likely that the Vanyukov furnaces were to be abandoned.[41] The NME was willing to accept such a major change in the project plans but insisted that the alternative technology be scrutinized anew to ensure that the environmental aspects would not be weakened.[42]

As it turned out, in February 2004 the Russians decided to stick with the Vanyukov furnaces after all. NIB declared, however, that the project would be delayed by at least two years. On NIB's request, the NME approved the necessary changes in the project plans and associated contracts that would reflect this delay. The termination date was moved from 30 June 2006 to 30 June 2008. The latter date, however, presupposed a test period for up to six months, so the actual delay would probably be closer to two and a half years.[43]

Although the uncertainties surrounding the choice of technology had cropped up due to circumstances beyond the control of Norilsk Nikel/Pechenganikel, when the designated buyer of large amounts of sulphuric acid changed plans, this episode did not bode well. Less than a year after the project had started up in earnest, the suitability of the single most important component of the modernization, the Vanyukov furnace which had been hailed as a superior solution only months earlier, had been called into question. This was an early sign that project plans, however meticulously composed, were by no means cast in stone. As such, that must have tempered any belief on the Norwegian side that Norilsk Nikel/Pechenganikel had finally reached a stage in their post-Soviet development where they could be considered a stable counterpart. In NIB's project reports the 'complex project organization structure in both Norilsk Nikel and Kola Mining [KGMK], leading to interfacing problems, including between the two Russian companies' was routinely listed as an important reason for delays.[44]

Re-organizations within the company structure, which to some extent must have reflected the still ongoing power struggle for oligarchic control of the increasingly profitable Norilsk Nikel concern, did not make for efficient project implementation.[45] There is, as we shall see in the following, much to suggest that the Norwegian donor was steadily losing faith in the project as time passed.

Reporting in mid-2005, NIB was still using the phrase 'satisfactory, though not as planned' to describe the progress of project implementation. However, that report otherwise makes for gloomy reading. Major difficulties had arisen in such key areas as the commissioning of the briquetting lines in Zapolyarnyi. This in turn led to problems in a pilot Vanyukov furnace in Monchegorsk that had been set up to prepare for the full-scale installation in Nikel. Due to insufficient supply of briquettes from Zapolyarnyi, the planned test runs at the pilot furnace could not be carried out. And this failure to produce test results meant that also other investments were postponed.[46]

Technologically, the modernization was proving exceedingly difficult to accomplish. However, the non-technical aspects were arguably equally demanding. The unstable situation within Norilsk Nikel/Pechenganikel, which led to uncertainty and unclear command lines, has been mentioned. In addition, the ever-changing conditions for a market-oriented enterprise like Norilsk Nikel/KGMK would constantly weigh in on assessments made by company executives. For Norilsk Nikel, the future company structure was still fluid. Possibilities on the Kola Peninsula ranged from substantial future investments in both mining and metallurgical operations, to dramatically reduced activity in both Nikel/Zapolyarnyi and Monchegorsk. Even a complete shutdown of Kola Peninsula factories was being contemplated.[47] The details of these complicated questions will not be discussed here. Suffice it to say that the modernization project was inevitably influenced by various decisions made by Norilsk Nikel/Pechenganikel managers in response to fluctuating market conditions. This was also reflected in NIB's 2005 report, where the uppermost reason for the many delays was described as 'alternative technical options arising from rationalisation of raw material production, supply and logistics, including a change in the market situation in the Northwest Region'.[48]

What was initially perceived as a strong point in the modernization effort from 2001 onwards – that the Russian side would be allowed to concentrate on its own technology and plan the project with a view to profitability while the Norwegian donors approved the plans on environmental merit – turned out to be a weakness. The attempt to unite in one project two different and at times opposing main objectives – profitability and environmental protection – proved very difficult. By mid-2005, it must have seemed increasingly clear to the Norwegian donors that their environmental demands would never be prioritized on the Russian side if these impinged on overarching industrial plans and thereby on potential future profits for Norilsk Nikel. Arguably, only if the environmental targets were reconcilable with the commercial logic of Norilsk Nikel would they stand a chance of being implemented.

The situation became even more acute when in the beginning of January 2006 NIB announced that modernization was lagging yet another two years behind schedule. The main obstacle to further progress this time was the recurrent problems involving the two briquetting lines in Zapolyarnyi. Claiming faulty deliveries, Pechenganikel was about to take the French provider of the new briquetting equipment to court. Obviously, settling that issue would take some time. There were also further glitches in the Vanyukov pilot furnace. All in all, NIB estimated, modernization could not be completed until the third quarter of 2010, a full four years later than the original plan. While assuring the NME of Norilsk Nikel/Pechenganikel's honest intentions of going through with the project, the bank asked yet again for permission to adjust all contracts according to the new time schedule.[49]

Unlike the first time when NIB had requested approval of a delay, the NME now considered the legal aspects of the required changes to project plans.[50] Basically, the ministry wanted to know whether the four-year postponement constituted a breach of the contract between it and NIB. In that case, the ministry could nullify its contract with NIB, which in turn would require NIB to annul its contracts with Norilsk Nikel and KGMK – and the whole modernization venture would be abandoned.[51] Arguably, the Norwegians had by then come to a point where they were looking for a graceful, and necessarily legal, way out of the modernization project, rather than retaining hopes of actually seeing it completed. However, the main concern at this point was to establish whether the delay constituted a violation of the contract.

Any thoughts of Norwegian withdrawal that may have existed proved premature, however. Legal experts at the office of the Attorney

General of Civil Affairs (*Regjeringsadvokaten*), when asked to assess the ministry's legal options, concluded that the delay that had been incurred was still within the limits set in the NIB agreement. As mentioned, this agreement established that the final deadline for project completion was 19 December 2011, exactly ten years after the agreement was signed.[52] Furthermore, as the attorney general considered the delays to be fully rational and therefore not in breach of the intentions of the agreement, there seemed to be no legal way out at this point. In concluding, however, the attorney general's office did point out that any future delays might well be in conflict with the modernization agreements, as these would probably exceed the ten-year provision and therefore give grounds for abandoning the modernization venture.[53] The ministry, after approving the delay, brought this last point to the attention of NIB,[54] making it clear that further delays would not be tolerated and that the NME had no intention of allowing the venture to go on indefinitely.

To sum up, the first five years of the most recent modernization effort did little to raise the level of trust between the two principal partners, Norilsk Nikel/Pechenganikel and the NME – or, more importantly, for Norwegian confidence in Russian industrialists. Repeated delays were nothing new in the modernization saga, but this time the disappointment at the lack of progress must have been even more keenly felt in Oslo. After all, the technical plans had been carefully composed and assessed, and the Russian firm had been allowed to develop these according to their own best ability and in their own best interest. In addition, when unspecified rumours of foul play by the Russians were repeatedly brought to the attention of the ministry,[55] this must have intensified a lingering Norwegian feeling of being duped. In light of these factors, the Norwegians decided to probe the legal framework that had been established in 2001/02, and carefully started to prepare an exit strategy. Hopes of actually completing an environmentally friendly modernization were fading, and plans for how to get out were taking shape.

## A carefully pursued divorce

The situation did not improve. By early 2007, the many uncertainties and questions that had riddled the Pechenganikel modernization since its early inception seemed as disputed as they had been almost 20 years earlier. At a meeting with the NME, NIB reported that questions of central importance to the modernization plans were being discussed once again in the Norilsk Nikel leadership. Contrary to what had been agreed upon, there was still uncertainty regarding which technology to use. The Vanyukov furnace was, apparently, still competing for prominence against the alternative and more modest refurbishment of the existing smelters and convertors. Of greater magnitude were the ongoing discussions about location. As several times earlier, Norilsk Nikel was considering the termination of smelting activities in Nikel and full transfer of ore processing to Severonikel in Monchegorsk.[56]

Obviously, the answers to these questions might rock the very foundations of the modernization endeavour. Another change of technology in Nikel would be problematic, although not impossible, for the involved Nordic donors and partners. However, any imminent termination of the processing plant in Nikel would render the whole project superfluous. The NME had, since discussing the matter in the 1990s, been adamant in its position: Norwegian funding would be directed only towards improvements at Pechenganikel, not Severonikel. Then there was the time factor. The Norwegians worried – or perhaps they hoped – that the decision-making process in Norilsk Nikel would take too long for the project to be completed within the time limits set in the agreements. While maintaining that time was about to run out for the project, the NME asked NIB to monitor the Russian decision-making process. NIB was also asked to set a last possible date for a Russian decision on location and technology if the project were to be completed without overstepping the time limit.[57]

Thus, the NME had in a sense commissioned a preliminary deadline. Instead of waiting for the project to peter out sometime in 2010/2011 due to a lack of decisions on the Russian side, the ministry wanted to establish the foundation for being able to claim that Norwegian support to the modernization, and consequently the whole modernization scheme, should in fact be terminated at an earlier stage. While this strategy had certainly matured in the ministry corridors over time, there is reason to believe that outside impetus, in the form of intensified press coverage of the slow pace of progress, helped spur the ministry on. From mid-February 2007 onwards, various reports and op-ed articles in Norwegian newspapers focused on the discouraging lack of progress in Pechenga. Many emphasized Norilsk Nikel's role as a giant on the international nickel and precious metals markets, and the concern's simultaneous disregard for environmental responsibilities. Was it – asked reporters, politicians and experts alike – right for the Norwegian government to grant a subsidy of 270 million NOK to an enterprise that boasted profits of almost 15 billion NOK in 2005?[58]

The rhetorical quality of this question relied on the obvious and increasing discrepancy between the Norwegian aid package and the fiscal situation of the Russian beneficiary. In the bigger picture, it reflected the economic boom that Russia had enjoyed since the turn of the millennium, and the lack of Norwegian policy adjustments in the same period. Norwegian policies towards its eastern neighbour, developed during Russia's catastrophic 1990s, were basically modelled as a comprehensive aid programme. With Russia's speedy revival the impact of Norwegian funding quickly decreased and gifts from the small state of Norway to the increasingly potent Russian power seemed obsolete.[59] In the case of Pechenganikel, this imbalance was particularly glaring. Owned by 'oligarchs' who were publicly flaunting their wealth, Norilsk Nikel/Pechenganikel was in the public discourse singled out as a Russian enterprise that in no way deserved a helping hand from the Norwegian government.

Evidently, this debate penetrated into the bureaucratic sphere as well. Citing recent questions 'in the media and elsewhere' about the effectiveness of 'spending money on projects in Russia', the Norwegian embassy in Moscow argued that the Ministry of the Environment could no longer leave the management of the modernization project to NIB alone. The time had come to reconnect directly with Russian authorities in the matter, the embassy maintained, although uncertainties abounded as to just what might be achieved:

> Presumably, the fact that Norwegian authorities get engaged might encourage the Russian side to consider the various options and alternatives that seem to be under consideration – alternatives that in sum convey an impression of substantial insecurity in Russia about future choices [concerning Pechenganikel].[60]

While presented as a constructive strategy, the embassy's proposal to engage Russian authorities anew reflected the mounting frustration with Norilsk Nikel. There was little to suggest that Russian authorities, who had been unable to control Norilsk Nikel while the concern was still state-owned, would be able to assert much influence now, some ten years after it had been privatized. However, as solid information directly from Norilsk Nikel was difficult to come by, liaising with the Russian government level must have seemed the sole viable option.

Yet another reminder of the gap between Russian commercial aspirations and Norwegian concern with the environmental aspects of the nickel industry came shortly thereafter, in April 2007. Norilsk Nikel and the governor of Murmansk oblast, Yuri Yevdokimov, went on a joint visit to Oslo. Though they met separately with representatives of the Ministry of the Environment, their messages were coordinated.[61]

Norilsk Nikel provided little new information about its internal strategy discussions and plans. The concern did seem intent on going through with the Vanyukov technology, but the question of Nikel's future as a company town was still undecided. What became clear in Norilsk Nikel's presentation was that the various agreements that the concern had entered with NIB (and indirectly with the NME) were all very much subordinate to overarching company decisions. Repeatedly, references were made to how implementation of measures under the modernization project – measures already agreed on in legal binding contracts – were 'under assessment' by the concern, rather than being implemented. This is made particularly clear in one sentence in the minutes from the meeting: 'When probed [the Norilsk Nikel representative] assured us that a modernization in line with the agreements with Norway is still a solution that is being considered.'[62]

Thus, the main objective for the Russian industrialists does not seem to have been to reassure the Norwegians – they did not conceal that they might withdraw from the whole modernization venture. Rather, the minutes indicate that their main objective was to sound out the Norwegian ministry for financial support for a full-fledged transfer of all processing activities to Monchegorsk. In presenting their plans, the Norilsk Nikel representatives emphasized both the industrial/commercial and the environmental benefits of moving the whole ore processing operation to Severonikel. Emissions in the Russian–Norwegian border area would be drastically reduced, and they also assured the ministry that their new operation at Severonikel would be 'the most environmentally-friendly in Europe.'[63]

When Governor Yevdokimov met with the ministry the next day, he was accompanied by the general director at KGMK. His message was basically the same as what had been presented by Norilsk Nikel: moving all smelting activities to Monchegorsk would be a good business solution for Norilsk Nikel and a good environmental solution for Norway. While stating that he was aware of the current Norwegian debate about financial support to projects in Russia, Yevdokimov underscored the need for Norway to keep on funding the modernization efforts, whether in Nikel or in Monchegorsk. The Norwegian reply was unequivocal: relocation to Monchegorsk might be a good environmental solution, provided that the smelter there was upgraded to handle the sulphur-rich briquettes from Zapolyarnyi. However, Norwegian support was aimed at the smelter in Nikel. If smelting there was terminated, no Norwegian support would be paid out.[64]

Thus, the Norwegians made it clear they were not about to be talked into a renegotiation of the project terms. Although they did acknowledge the apparent environmental benefit to the border area if ore processing were to be moved further south and east on the Kola Peninsula, they chose to stand firm on the assessment made by the Norwegian Pollution Control Authority in 2000, in which the possibility of funding a modernization project in Monchegorsk had been rejected outright.[65] The Norwegian resolve to avoid yet another round of see-sawing in the seemingly endless modernization saga had grown firm.

This resolve must have been further strengthened a few weeks later. NIB, having assessed the contractual framework that surrounded the modernization project, gave the Russian industrialists less than a year to make a final decision on both technology and location of smelting activities. According to the bank, all necessary decisions and agreements would have to be made by the first quarter of 2008. If not, Norilsk Nikel would not be able to complete the modernization on time and would consequently be defaulting on the financing agreements made with NIB.[66] In other words, only a short year's wait would now be enough for the whole modernization project to be rendered null and void.

In July 2007, an episode took place that briefly increased the domestic pressure on Norwegian authorities. Due to specific weather conditions, sulphur dioxide concentrations in Pechenga and neighbouring areas in Norway were extremely high. In Nikel, levels were reported to be twenty times above normal, and children were allegedly kept indoors for fear of contracting pulmonary diseases. On the Norwegian side of the border, in Kirkenes, there were even attempts to reinvigorate the 'Stop the Death Clouds' campaign. One MP from Finnmark County raged against the Norwegian government for failing to inform the local population sufficiently about the health hazards in the border area.[67] Although this episode literally blew over quickly, it kept the issue of airborne pollution from Nikel alive in the public discourse – and Pechenganikel remained a constant irritant in Norwegian–Russian relations.

After this, the modernization saga entered a lull. Almost a whole year passed before NIB in June 2008 confirmed that there remained no possibility of timely completion of the modernization project. At this prompt, even the Norilsk Nikel management agreed to NIB's assessment of the situation. Thus, default had become a fact – a fact which would have allowed the NME to suspend disbursements of funding and declare publicly that Norilsk Nikel and KGMK were in default.[68] Interestingly, the ministry opted to avoid rocking the boat. All payments were suspended, but without any public commotion. The main reason was the uncertainty regarding the future processing of local ore on the Kola Peninsula. Norilsk Nikel had yet to decide whether to terminate in Nikel and transfer all ore to Monchegorsk, or to keep up activity at Pechenganikel's polluting smelter.

The first part of the modernization – the briquetting line in Zapolyarnyi – was about to reach completion, and continued activity in the still-unchanged smelter in Nikel was considered highly problematic by the Norwegians. As described in this chapter, the briquetting process was designed to eliminate emissions in Zapolyarnyi by not releasing sulphur from the compressed ore. However, this would provide environmental benefits only if the briquettes were further processed in a modernized smelter. As things stood, the sulphur that had been contained in Zapolyarnyi would be transported further west to Nikel, where it would be released in the unchanged smelting process. Paradoxically, then, the result of the halfway modernization would be more, and not less, sulphur dioxide in the air over Norwegian territory. Aware of this situation, the NME wanted to retain any possible leverage on Norilsk Nikel/KGMK they had, and therefore decided to stay officially committed to the agreement with NIB.[69]

Any lingering hopes of an imminent decision by Norilsk Nikel on the future structure of its Kola Peninsula operations were thoroughly crushed by the financial crisis that started in the fall 2008. As an export-oriented concern, Norilsk Nikel was extremely vulnerable to fluctuating price levels on the international raw materials markets. By February 2009, the price of nickel had plummeted from its peak in 2008 at USD 75,000 per ton to a mere USD 9,000 per ton. The deteriorating situation in the concern led Russian state-controlled banks to buy shares and demand representation in the Norilsk Nikel boardroom. There ensued several changes in the concern management, making the situation even more uncertain.[70]

Confronted with this information, the NME still stood by its commitment to the modernization project, at least nominally. Again, the underlying rationale was to retain some leverage over Norilsk Nikel. An additional motivation was that the ministry wished to avoid being seen by the Norwegian public as responsible for continued and even increased pollution from Pechenganikel. For these reasons, the Norwegian side declared that 'the agreements should be upheld until the Russians withdraw or until the contracts expire.'[71] Evidently, then, at this point the end game was well under way. There were no longer any real ambitions on either side of actually going through with the modernization. What remained was tactical manoeuvring to minimize the repercussions of the divorce.

The contracts were unilaterally cancelled before they expired. A financially weakened Norilsk Nikel formally withdrew from the project in October 2009, apparently to avoid having to pay further management fees to NIB.[72] With that move, the modernization project was terminated for all practical purposes. Nevertheless, the Norwegians still resisted a formal withdrawal. Even at this late point they wanted to be able to appear blameless, and waited until March 2010, when all deadlines in the project plans had been unequivocally overstepped, before formally pulling out.

Together with various government agencies, the NME discussed how best to respond to Norilsk Nikel's unilateral withdrawal. Having assessed and rejected the option of outright passivity, they concluded that the choice stood between what they referred to as the 'pro-active' and the 'reactive' approaches. The 'pro-active' option entailed re-negotiating all agreements with a view to achieving some environmental improvements at Pechenganikel,[73] albeit only those which could be implemented at minimal cost. The main idea behind this solution was to use whatever leverage Norway had over Norilsk Nikel; if the concern agreed, they would not have to face Norwegian demands for reimbursement of previously disbursed funding. However, the environmental benefits to be derived from such an approach were deemed to be miniscule. The 'reactive' option involved activation of the paragraphs in the agreements that entitled Norway to full reimbursement of all its investments in the incomplete modernization of Pechenganikel.

After a thorough examination of legal aspects,[74] the Norwegians opted for the 'reactive' approach. Having spent close to 48 million NOK on the modernization project, the Ministry of the Environment demanded that Norilsk Nikel pay back, via NIB, all Norwegian funds. Norilsk Nikel now found itself between a rock and a hard place. On the one hand, the Norwegians threatened to claim international default if the funds were not reimbursed. That would obviously have negative implications for the industrial and fiscal credibility of Norilsk Nikel and might even lead to an international blacklisting. On the other hand, Norilsk Nikel's image was also likely to be damaged if they did pay back the money. Full reimbursement would not only confirm the failure of Norilsk Nikel/Pechenganikel in the modernization as such, but also demonstrate to the outside world that the Russian nickel industry was uninterested in, or unable to effect, actual improvements in its environmental record.

Norilsk Nikel had to make a choice, however unpleasant the options might be. As it turned out, the concern apparently wanted to be free of all commitments to the Norwegians and to end the modernization saga once and for all. On 14 September 2010, Norilsk Nikel agreed to reimburse 46,677,670.73 NOK, to be transferred to NIB and then to the NME.[75] The concern, however, set one condition: that the Norwegians kept this settlement to themselves and refrained from a disclosure of the deal.

Thus, almost exactly twenty years after Norwegian prime minister Jan P. Syse proudly proclaimed his government's intention to support pollution control in Pechenga and thereby 'save the natural environment in Finnmark', the Norwegian subsidy had been unequivocally withdrawn. But the problem was still there, very much so. Even if emissions from Pechenganikel had, for various reasons, decreased since the early 1990s, sulphur dioxide discharges and heavy-metal pollution were still a source of great concern in the Norwegian–Russian border areas.

## Back to square one?[76]

Although the modernization project had finally been put to rest, the Norwegian government, most notably the Ministry of the Environment, had no intentions of abandoning the Pechenganikel issue altogether. The situation in the summer of 2007, when high levels of sulphur dioxide in Nikel and Sør-Varanger across the border in Norway created a heated debate (see earlier in this chapter), had fully demonstrated that Pechenganikel's smelters still represented a major threat to the natural environment and, according to some, to human health as well. Information that the NME had received from Norilsk Nikel indicated that the ore reserves could be producing for at least another sixty years.[77] There was no reason to believe that the problem would go away by itself.

The nickel plant seen from the east, 2008. Note the extensive emission of smoke from the smelter building and attempts at revegation in the forefront.
Photo: Anne Berteig

Emissions to air

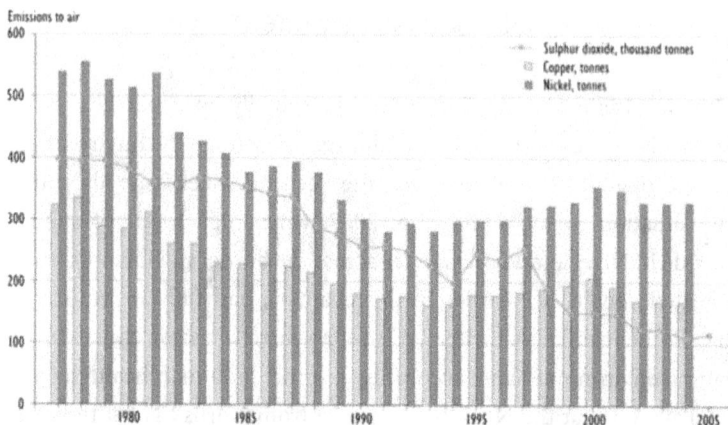

**Figure 1** Emissions of sulphur dioxide, copper and nickel to air from the Pechenganikel enterprise, 1977–2004.
Source: Office of the Finnmark County Governor (2008), p. 7.

Fresh data confirmed that average emission levels from Pechenganikel, although substantially decreased since the peak of the late 1970s, were still very high. In 2008, the final report from the 'Pasvik Programme', a comprehensive collaboration involving regional environmental authorities in Norway, Finland and Russia, was published. Commencing in 2003, this program aimed at monitoring the anticipated environmental effects of the modernization project. It also compiled extensive data series and traced long-term developments in the border area. The main finding was that sulphur dioxide levels, which had hovered at about 400,000 tons annually in the late 1970s, had stabilized at slightly above 100,000 tons by 2004 (see figure 1).

Though this represented a 75 per cent decrease, Pechenganikel alone was still discharging annually about five times as much sulphur dioxide as all Norwegian sources combined.[78] Even more worrying were the emissions of heavy metals. Dust particles containing nickel, copper, cobalt and other toxic metals were being released in large amounts from the various production processes at Pechenganikel. Unlike the decrease in sulphur dioxide emissions, the discharges had been increasing with some of the heavy metals. Damage was especially evident in the adjacent Pasvik watercourse and its aquatic flora and fauna.[79]

Perhaps the most troubling fact behind these figures was that any environmental improvements, such as the decrease in sulphur dioxide emissions, were barely the result of active measures taken to combat pollution. Imported Norilsk ore from Northern Siberia had been the main culprit behind Pechenganikel's enormous increase in emissions in the 1970s. The gradual curtailing of this ore at Pechenganikel due to the high costs associated with sea and rail transport throughout the 1980s and 1990s was by far the single most important reason for lower emission levels. The worrying tendencies in the heavy metals emissions could only serve to reinforce the impression of an industrial company that accorded little or no weight to limiting the negative impacts of its production. In 2009, the Norwegian authorities declared that the emissions from Pechenganikel were 'unacceptably large', especially the heavy metals discharges.[80]

As noted earlier in this chapter, when the modernization project was terminated in 2010, there was every reason to believe that sulphur dioxide emissions from Nikel would increase soon. The briquetting line in Zapolyarnyi, then nearing completion, would soon start feeding the smelter in Nikel, and discharge figures were expected to escalate. Basically, the 45,000 tons of sulphur dioxide per year contained in Zapolyarnyi would be moved with the sulphur-rich briquettes to Nikel, and then released into the atmosphere even closer to Norway. When confronted with questions from the press about this in 2009, the NME summed up the problem and briefly outlined its approach to the issue:

> We will continue to say loudly that the situation in Nikel is unacceptable. The problem lies first and foremost with the company, which can afford to clean up its emissions, but fails to do so. But also, federal Russian authorities carry a big responsibility for not being able to put strict enough demands on the enterprise.[81]

With this, the ministry signalled a different line of action. No longer was Norilsk Nikel the main addressee for Norwegian complaints, and no longer would the company be offered carrots in the form of environmental subsidies from Norway. Now the spotlight would be re-directed to the level of the Russian authorities, as had been the case during the final years of the Soviet Union's existence and the first post-Soviet period when Norilsk Nikel and Pechenganikel were still state-owned enterprises.

Contact on the ministerial level had occurred before the modernization was formally abandoned, albeit after it had become publicly known that the project had failed. It was in fact the deputy minister in the Russian Ministry of Natural Resources who took the initiative to meet with State Secretary (junior minister) Heidi Sørensen of the NME in order to discuss pollution in the Pasvik valley in late February 2010. Apparently, the Russian authorities were also interested in the Pechenganikel question, and their initiative was interpreted by the Norwegians as an unusual attempt to be forthcoming in the matter. However, the meeting proved to be anything but a demonstration of bilateral concord. Though both parties emphasized that emissions from Pechenganikel were excessive, there was a marked discrepancy about which emission levels each side considered acceptable. Pointing out that sulphur dioxide discharges from Pechenganikel had already been reduced substantially, the Russians stated that it would be difficult to force Pechenganikel emissions to less than 40,000–45,000 tons sulphur dioxide annually, which would be in line with Russian regulations. The Norwegians referred to the modernization agreement, which stipulated a reduction to 12,000 tons sulphur dioxide annually and maintained that this was the maximum level that the natural environment could absorb without sustaining significant damage.[82]

The minutes from the meeting testify to a high level of tension between the two parties. The Russian ambassador to Norway Sergei Andreyev, who was also present, stated in no uncertain terms that the Norwegian calls for emission reductions were simply unrealistic. Pollution, they said, was not the sole worry of the Russian authorities: they also had to safeguard the financial wellbeing of the concern and ultimately protect employment and local settlements. When the Norwegians indicated that Russia should bring its emission regulations in line with 'a level that the environment can actually tolerate', the reply was terse: 'Other countries' regulations and wishes cannot define which laws shall be applied in Russia.' To this, the Norwegian side replied, equally curtly: 'In Norway [it is] normal to be inspired by other countries' regulations.' Junior minister Sørensen encouraged Russia to 'establish an equivalent practice'.[83]

As we see, the gap remained between the two countries' approach to environmental politics as it had been in the late 1980s. The parties simply did not see eye to eye when it came to environmental protection from polluting industries. Nevertheless, there was one common denominator: the current level of emissions at Pechenganikel was unacceptably high, and something should be done about it. The question was just what needed to be done. The Russians wanted to support the ongoing process in Norilsk Nikel to determine a new technological concept for the Pechenganikel smelter. The Norwegians, hesitant to the idea of yet another project development, pointed out there already existed a technological solution, from an earlier round.[84] In any case, it became obvious from statements at the meeting that any serious Russian action taken to tighten Russia's environmental regulations would have to be preceded by an 'unequivocal signal from the very top'.[85] The industrial driving forces that worked against stricter environmental legislation, it was said, were too powerful for the Russian Ministry of Natural Resources to take on without the backing of the president of the country.

In a sense, this meeting pointed towards Norway's current strategy in the Pechenganikel issue. In April 2010, President Dmitri Medvedev was scheduled to visit Norwegian prime minister Jens Stoltenberg in Oslo. The prime minister brought the issue to the attention of the Russian president.[86] Although Medvedev's visit will primarily be remembered as the occasion when the Norwegian and Russian leaders finally agreed on the long-disputed delimitation line in the Barents Sea, the resultant joint declaration also contained a brief passage on the Pechenganikel issue:

> The parties agree that emissions from the nickel production in the Pechenga region in Murmansk oblast give cause for concern and must be brought down to a level that does not harm health and environment in the border area. The Russian side will contribute to implementation of necessary measures to reduce the emissions. The parties agree to cooperate to gather objective information about emission levels, strengthen the future [pollution] control and to build up environmental monitoring in the area.[87]

Although these few lines did not figure prominently in the eight-page declaration, they did give the NME a starting point. The ministry could now argue that, with Medvedev's and Stoltenberg's statement, the Russian Ministry of Natural Resources and Ecology had in fact received the authoritative signal it had claimed would be necessary for it to make any headway vis-à-vis Russian industry.

\*   \*   \*

After finally abandoning the subsidization scheme in 2010, Norwegian authorities have basically worked to impress upon their counterparts in Moscow that the issue of sulphur dioxide emissions from Pechenganikel is a responsibility that weighs heavily on the Russian state, and that failure to deal effectively with it will place political strains on relations between the two countries. The Norwegians refrain from putting direct pressure on the Norilsk Nikel concern or KGMK, instead keeping the issue within the framework of their bilateral relationship with Russia. Seeking to couple the Pechenganikel emissions with other policy areas, such as trade, the Norwegians have brought the matter up in various arenas. This work has yielded some results. In early 2011, the Norwegian–Russian Governmental Commission on Economic, Industrial and Scientific-Technical cooperation decided to set up a bilateral working group. Its mandate was to be developed in concert by the Russian Ministry of Natural Resources and Ecology and the NME. Due to obstruction on part of Norilsk Nikel – the company has simply not turned up at meetings where its presence has been required – it was difficult to reach binding agreements. The approach, however, has been to seek a common understanding of what the environmental issue in Pechenganikel represents, to agree on upper emissions limits and, finally, to define guidelines for technological solutions to the problem.

In a sense we are indeed back to square one. Almost thirty years after Pechenganikel emissions first became a major issue in bilateral relations between the Soviet Union/Russia and Norway, the parties are still grappling with very fundamental questions. Again, it has become a matter for the governments to resolve. Most remarkably, even after all this time, there is still a large gap between the two parties in the basic understanding of the problem.

# 6

# Conclusion

Perhaps the most puzzling point in the history of the Pechenganikel modernization saga is the glaring lack of results. Reading through the documentation, one cannot fail to be baffled by the inability of the two parties to get any closer to alleviating the environmental problems in the Pasvik valley. One may argue, as NME officials have in conversations with the author, that the modernization project fell victim to historical circumstances that rendered implementation extremely difficult. There is no doubt that Russia's transitional period, involving two major financial breakdowns, highly controversial privatization processes and, less apparent, widespread corruption, made for demanding conditions. However, these factors were part and parcel of what made the whole project viable in the first place – the Soviet collapse. More important, I would argue, the sheer number of hours (years) spent on moving the project forward does lead to an expectation of some results, either on the ground or in the less tangible sphere of a common understanding. As will become apparent in the following, the two parties were as distant from each other in 2010, and in some respects even more distant, than they were in 1990.

Equally surprising as the lack of results are the apparently endless persistence and stamina on both sides. Even in the face of numerous disappointments, the NME kept up hopes of eventual success. Similarly, Norilsk Nikel/ Pechenganikel agreed to the Norwegian approaches for a while, albeit with varying levels of commitment. And yet, the desired results failed to emerge. Why, then, in the face of repeated failure, was the Pechenganikel modernization not simply demoted from an active and ongoing collaborative project? Could it have been better dealt with as one of many bilateral Russian–Norwegian issues that were regularly discussed in meetings between the representatives of the two states? After all, progress was basically non-existent, and traditional diplomacy could hardly do worse.

It was primarily the Norwegians who kept the modernization project alive through their stubborn insistence that Norilsk Nikel/ Pechenganikel and the Russian Federation had a moral responsibility to clean up the mess left after decades of environmentally unregulated nickel production. Moreover, according to the Norwegians, such a clean-up would be possible because of the funding that they would provide. As in so many foreign policy issues, the Norwegian consensus in the Pechenganikel case seems quite amazing: not one of the six Norwegian governments of varying political persuasions that held office between November 1990, when Jan P. Syse's conservative coalition was dissolved, and 2005, when Jens Stoltenberg's red–green coalition government won the election, saw fit to change the approach to the Pechenganikel problem. Only in 2010, after two decades of unsuccessful and exhausting attempts to cajole the Russians into curtailing their pollution, were further efforts along the same lines of dealing primarily with the Norilsk Nikel concern and Pechenganikel pronounced futile. Having finally severed all direct ties to Norilsk Nikel, the Norwegian government stepped up its political pressure and gave more attention to the Pechenganikel pollution problem in bilateral talks with Russia. We will investigate the factors that may explain Norway's dogged persistence, after first examining some theories as to motives on the Russian side.

## Motives in a fruitless process

In his 2001 *Smokestack Diplomacy*, Robert G. Darst offers a comprehensive and bold interpretation of the underlying factors motivating Russian authorities and industrialists when, after the Soviet breakdown, they were faced with the prospects of foreign funding for environmental clean-up. According to Darst, the Russians quickly came to see domestic pollution as an asset rather than a liability, especially if it happened to affect the territory of an affluent neighbour. Promises of foreign funding, he argues, created a dynamic in which the Russians took the position of the extortionist:

> Why close an antiquated, environmentally threatening facility – or take any steps to clean up its emissions – if there is a reasonable chance that its very 'dirtiness' will lead to transnational bailout? For a power plant or factory in the former socialist bloc, few assets were more valuable than the ability to pose a serious transboundary environmental threat to one or more of the affluent capitalist states, for this threat brought with it the prospect of externally financed modernization.[1]

Thus, he maintains, Western states wanting to subsidize pollution reduction schemes in Russia were in fact subjected to what he calls 'environmental blackmail'.[2]

Darst applies his theory to a series of cases, the Pechenganikel modernization among them. In discussing the Pechenganikel case, he explains why, in his opinion, the project did not come to fruition: it was never economically viable if the local ore resources in Pechenga would soon be exhausted anyway, leading willy-nilly to the demise of Pechenganikel. Unlike the Russian state, the private Russian owners who entered from the mid-1990s were able to see this clearly. That the modernization was contemplated for as long as it was, Darst argues, was first and foremost a result of the economic interests of the Nordic entrepreneurs involved.[3] Motivated by their potential gains from the project, Norwegian, Finnish and Swedish contractors prolonged Norway's involvement through intense lobbying in Oslo.

Although Darst, somewhat surprisingly, leaves his readers in the dark as to whether he thinks 'environmental blackmail' was actively applied against Norwegian donors in the Pechenganikel project, his emphasis on *Nordic* entrepreneurs would suggest that it was in fact not. A situation where the main driving forces behind continued modernization efforts were Norwegian, Finnish and Swedish industrialists seems to exclude the possibility of Russia acting as a shrewd and actively shot-calling extortionist. However, and this seems to be Darst's argument, the Russians were conducting *passive* blackmail of Norway by doing nothing at all about the pollution problem in Pechenga. In the hope of receiving subsidies, the Russians remained willing to negotiate with and interact with Nordic entrepreneurs and authorities – while, one can surmise from Darst's basic premise, resisting solutions to the environmental problems at hand.

Though Darst's emphasis on Nordic commercial interests is relevant for what happened during the 1990s, it cannot account for the final ten years of the modernization saga. Writing at the turn of the century, Darst could of course not have the benefit of knowing what happened later, when the private owners of Norilsk Nikel actively promoted a modernization of the Pechenganikel smelter based on their own technology. It can hardly be argued that, at this point, the sidelined Nordic entrepreneurs played any part in prolonging the project. By then, Norilsk Nikel had calculated that the prospects for the concern's Kola operations were bright enough to merit some investment. Understandably, here the concern preferred to curtail its own expenses by activating Norwegian funding for an upgrade. The Russians were able to convince the NME to maintain its economic support. That the concern later tried to transfer the modernization to Severonikel in Monchegorsk does not fundamentally change this premise: the ore would still be mined in Pechenga, and a smelting plant would still be in operation on the Kola Peninsula. If the upgrade, which the Russians were contemplating for reasons of efficiency, could also satisfy Norwegian environmental demands, there was no reason for Norilsk Nikel not to accept the gift.

Another search for Russian motives has yielded results similar to Darst's interpretation of the Russians as extortionists. In her 2005 thesis on the Pechenganikel project Gro Elisabeth Olsen identifies what she calls the Russian 'tactics of the vise'. In short, Olsen demonstrates how the Russians repeatedly made last-minute changes to important factors related to imminent agreements with the Norwegians, thereby making postponements necessary. Having first given the impression that they were on board with their optimistic Western counterparts, the Russians would suddenly shift position, giving rise to an atmosphere of deep consternation. And then, when the Russians again lifted the pressure, or in Olsen's phrase 'released the vise', the Norwegians would be willing to go the extra distance to find a solution. These tactics, according to Olsen, were not only meant to yield better negotiation results for the Russians; they also prolonged the modernization project.[4]

While Olsen's points are well argued, her imagery of a vice that is tightened and relaxed seems to presuppose that the various Russian entities involved in the Pechenganikel modernization together formed one unitary rational actor seeking to maximize the benefits associated with the negotiations. If, however, we consider the instability in Russian society at large as well as in the industry during the 'state of emergency' in the 1990s, this can hardly be said to be the case. The Russian scene was extremely fragmented, responsibilities were unclear, and power balances were fragile. This also applies to the situation within the Norilsk Nikel concern. Instead of acting as one unit, the different Russian actors sent mixed messages that often seemed contradictory. While this, in Olsen's view, would be a sign of the Russians applying the 'tactics of the vise', that image is probably a more accurate description of how things were felt on the Norwegian side than how they actually worked in Russia.

That said, this does not mean that the various Russian players did not have their own self-serving agendas. Both Darst's and Olsen's theories merit attention, and it does seem likely that, in certain situations, aspects of 'environmental blackmail' and 'tactics of the vice' were applied to achieve specific short-term objectives. However, it is hard to imagine that such methods were applied in a sustained and coordinated manner on the top managerial level in Norilsk Nikel or in the relevant Russian ministries. Throughout most of the 1990s, the manifold Russian interests that were involved were simply too disorganized, divided by conflicting aims. Thus, it is more likely that the mixed and often confusing messages from Norilsk Nikel, Pechenganikel and various Russian ministries were expressions of a fragmented Russia – and not of a well-coordinated counterpart that was trying to shake Norway down for money.

When circumstances had changed by the end of the 1990s and a coordinated and consolidated Norilsk Nikel introduced the Vanyukov technology, the contracts between Norilsk Nikel, KGMK, NIB and the NME left little room for blackmail or elaborate negotiation techniques. Although reading backwards is a questionable activity and intent may differ from consequence, we may conclude that if the Russians in fact did try to squeeze money out of the Norwegian coffers, they failed miserably. As we have seen, the 30 million NOK that was spent went to the Norwegian–Swedish consortium PRC. And everything transferred to Norilsk Nikel, almost 47 million NOK after 2000, was eventually returned to the NME in 2010.

Turning to the other side of the Pechenganikel modernization saga, we find a set of Norwegian and Nordic actors that may have been better coordinated than the Russians, but they were no less diverse. While the unsuccessful outcome of the modernization efforts must be ascribed to all the parties that were involved, there is little doubt that the most enduring interest in the project was Norwegian. Although initially discussed as a commercially motivated business venture between Soviet and Finnish authorities, the Pechenganikel modernization became presented as first and foremost an environmental effort once the Norwegians entered the stage in 1990. After the Finnish government showed significantly less interest following the rejection of Outokumpu's plans in 1992, the project was fully driven by the Norwegian government, and by extension its Ministry of the Environment. The Norwegians held the key to prolonging the Pechenganikel project, through the promise of substantial subsidies. To find an explanation as to why the modernization efforts could meander on for two decades with no positive results, then, we must look to Norway.

The basis for the modernization effort was always the environmental situation that had arisen in the Pasvik valley, and the 'death clouds' that were rolling menacingly over Norwegian territory. Environmental advocates played a central part in shaping Norwegian approaches to the Pechenganikel emissions, especially in the early stages. Initially, popular environmental demands, channelled primarily through the 'Stop the Death Clouds' campaign in Sør-Varanger, were instrumental in putting increased pressure on the government. This campaign, with its depiction of a grim future for the natural environment of North Norway unless decisive action were taken, proved exceedingly effective.

Although this ad hoc movement quickly faded after achieving its primary goal of provoking Norwegian state commitment, the more established environmental movement continued to keep tabs on the modernization project. Among the more active in this respect was the Bellona Foundation. In 2010, this organization published a report on Norilsk Nikel's environmental record and was granted funding from the Norwegian authorities to hold a large public meeting in Nikel that was also attended by Norilsk Nikel/KGMK representatives.[5] Interestingly, Bellona had recommended already at an early stage that the Norwegian authorities abandon their subsidization strategy in the Pechenganikel case, and direct the money towards other environmental target areas.[6]

That said, there was much more to the Pechenganikel modernization than just improving the environment. As shown in previous chapters, and as pointed out by Darst, Nordic entrepreneurs – most notably, the Norwegian companies Kværner Engineering and Elkem Technology – were central in defining the content of the modernization project throughout the 1990s. Until 2000, when Norilsk Nikel insisted on using its own technology, the very same companies were the main supplier of information about progress in technical negotiations to Norwegian authorities. Thus, a party that had a strong vested interest in the continuation of the modernization efforts was also providing much of the information that helped prolong the project. As these entrepreneurs were driven by considerations of the profits they might and in fact did accrue from participating in the modernization venture, there would always be a danger of over-optimism in their reports to the NME. Thus, they might have contributed, as Darst maintains, to keeping the

Norwegian authorities bound to the project longer than what otherwise would have been the case.

The patchwork of various commercially motivated interest groups on the Norwegian side was even more complicated. Locally in Norway, and most notably in the northerly areas bordering on Russia, there were great expectations of spin-off effects from the modernization project. Local contractors in Finnmark county were lobbying Norwegian authorities for a share in what they hoped would become a major industrial undertaking in an area that needed all the employment it could get. This was very much the case in Sør-Varanger, where the gradual downscaling of partly state-owned AS Sydvaranger culminated in 1996 with a shutdown of the iron works. This created fertile ground for discontent with the central government in Oslo, which northerners had long felt had been neglecting the periphery. When the modernization plans came up, local entrepreneurs seized the opportunity to demand subcontracts that would benefit them. This meant that the Pechenganikel modernization was not only a foreign policy concern for the Norwegian authorities' it also touched on long-standing centre–periphery tensions in the country's domestic politics. By claiming what they saw as their rightful share of a partly Norwegian-funded modernization project – a claim that was politically difficult to brush aside – northern Norwegian economic interest groups also contributed to complicate and prolong the Pechenganikel modernization negotiations.

But then, as mentioned in this chapter, the Nordic entrepreneurs vanished after the introduction of the Vanyukov technology in 2000 (except for Boliden Contech, in a peripheral position). Their role in prolonging the modernization efforts can therefore account for only one of the two decades of the saga. After 2000, the Norwegian authorities employed two complementary approaches to their involvement in the Pechenganikel modernization venture. Firstly, and most importantly, the Ministry of the Environment insisted that a comprehensive contractual framework be developed that would safeguard Norwegian interests and provide for the possibility of withdrawal if this became necessary. Secondly, the Norwegian authorities made decisions based on the regular reports from NIB, but there is no reason to believe that these reports were inaccurate or overly optimistic.

Rather, in this second decade of the modernization saga, the longevity of the process can be explained by primarily contractual stipulations. The contracts not only protected Norwegian interests, but also stated that the agreements were to remain valid for a period of ten years. Norway had no wish to be blamed for the failure of the project by stepping out before this period was over and therefore remained outwardly committed to the project. This commitment was scarcely a reflection of Norwegian belief in the eventual success of the modernization. As we have seen, the Ministry of the Environment were hedging their bets, and arguably started preparing the way out, already in 2006.

## Irreconcilable world views

As shown in Chapter 3, it was clear from the very first bilateral environmental meeting in Finnmark in 1986 that Norwegian and Soviet officials differed in their perception of and approach to the pollution problems in the Pasvik valley. These divergences continued throughout the 1990s and, I would argue, persist in Russian–Norwegian relations today. Before turning to the specific ideas and world views that underpinned the differences between the two cultures that clashed in the Pechenganikel modernization saga, let us look at the bilateral context that surrounded the project.

With the Soviet collapse in 1991, Russia was thrown into a deep recession. Norwegian tendencies to view the Soviet Union as a developing country that needed help had emerged already in the late 1980s and evolved into official policy when the Union disintegrated. From then on, the Norwegian discourse on Russia featured descriptions of social catastrophe, political chaos and environmental disaster. True to Norwegian tradition, an apparatus – comprising both state-sponsored non-governmental organizations and the state bureaucracy itself – was set in motion to provide foreign aid to Russia. Ambitious plans to replace the Soviet legacy with Western values were declared. Unsurprisingly, the Norwegian contribution to a restructuring of Russian society was scarcely palpable. However, the sincere naivety of this ambition bears witness to how much Norwegian authorities actually thought they could achieve in the first ten years of the post-Soviet era.

In line with this new image of Russia as a developing country, there was an increasingly strong propensity among Norwegian politicians and bureaucratic decision makers to disregard the traditional asymmetry in Soviet–Norwegian bilateral relations. With a feeble-looking Russia emerging from behind the crumbling Iron Curtain, the small state–great power imbalance was in some quarters of Norwegian politics seemingly inverted: now the time had come for Norway to call the shots. Much of this was related to the fact that throughout the 1990s, the Russian economy was tottering on the brink of bankruptcy. Consequently, all bilateral projects would have to be funded by Norwegian sources. Though the authorities steadfastly labelled the Russian–Norwegian interface 'cooperation', the Norwegian way of talking about Russian affairs became coloured by the de facto donor–recipient relationship between the two states.

After the immediate post-Soviet period, when many in Russian politics, media and the population at large euphorically embraced everything Western – including the utter rejection of the Soviet legacy – this set of Norwegian assumptions became an impediment in the bilateral relationship. By the mid-1990s, the discrepancy between the Norwegian understanding of the post-Soviet area as an inferior developing economy and the evolving Russian self-perceptions had become acute. A growing distrust of Western intentions was emerging in Russian society. As the chaotic transitional period during Yeltsin's presidency unfolded, fewer and fewer embraced Western liberal ideals, and some even viewed their introduction in Russia as part of a covert Western warfare strategy. Furthermore, any suggestions of an inversion of the small state–great power asymmetry in Russian–Norwegian relations would meet with little understanding in Russia. As citizens of one of the world's largest countries, many Russians reacted negatively to Norway's 'helping hand'.

Moving on to the specifics of the Pechenganikel modernization, we see a parallel pattern emerging. The Russians were confronted with Norwegian pressure in the form of a gracious gift package – combined with the morally charged expectation that they 'clean up their own mess'. The Russian authorities were obliged to respond. But what exactly did they have to respond to?

Firstly, there was the Norwegian perception of the Kola Peninsula as an environmental disaster zone. This image resonated poorly in Russia. Though very aware of local pollution problems, the residents of Murmansk *oblast* – journalists, politicians and civil servants alike – would tend to emphasize the abundance of untouched nature on the peninsula. Secondly, the Norwegian disregard of Russian (and previously Soviet) environmental expertise implicitly expressed through 'capacity-building' efforts – programmes that aimed to export Norwegian environmental management practices to Russia –[7] must have irked the pride of Russian professionals. Norway's bureaucrats, like most of the Western world, were largely convinced that the Soviet environmental legacy was one of despondency beyond despair. According to Jonathan Oldfield – and I would be inclined to agree with him – this conviction was accompanied in the West (most likely also in Norway) with a certain sense of triumph:[8] not only had the socialist camp lost the Cold War fight for ideological supremacy – it had wrecked its natural environment in the attempt to compete. A logical continuation of this view was the assumption that Soviet environmental awareness was non-existent.

These diverging Russian and Norwegian outlooks constituted only part of the challenging framework surrounding the Pechenganikel modernization efforts. Perhaps the deepest chasm between the two sides was to be found in their approaches to environmental protection. On the one hand, there was the Norwegian ecological movement that had emerged via classical conservationism to present-day environmentalism, mainly focused on the degrading effects of industrial pollution on ecosystems. The establishment of the NME in 1972 was the result of this environmentalism maturing into a widely accepted reaction to post-war industrialism.

As discussed in Chapters 1 and 3, there was never a parallel development in the Soviet Union. Here, industrial pollution was an undesirable, but tolerable, side-effect of the all-important activity of the many Soviet manufacturing enterprises. To the extent that environmental improvements might occur, this would be the result of technological advances developed within the industry itself. Achieving the goal of 'complex utilization' of natural resources would both raise efficiency and curtail unwanted discharges. Thus, industrial pollution was in the Soviet tradition a matter for engineers, not for political agencies or popular movements. Institutionalized throughout the Soviet period, this approach was adopted in post-Soviet Russia as well.

Clearly, then, the success of the Pechenganikel modernization hinged largely on an approximation of diverging ideas. Let us revisit the model dealing with world views and foreign policies, introduced in Chapter 1. As noted there, sets of ideas that shape collective outlooks and thus also foreign policies can be organized in a three-level hierarchy of *world view* (the overarching set of values dominant among a given group of people), *principled beliefs* (shared normative ideas of that group) and *causal beliefs* (commonly accepted cause–effect relationships).

Applied to the Pechenganikel modernization, these categories come to life. The Norwegian *world view*, shared with most Western European countries, is rooted in post-war capitalist liberalism in the reverence of such ideals as the rule of law, individual freedom and human rights. The Russian world view was derived from a quite different tradition – one that prioritized collective goals above individual aspirations and rights. In 1991, Russian society had recently seen how the very foundation of its long-standing value set – Soviet socialism aimed at the collective goal of a truly communist, classless society – crumbled to pieces. Not yet having replaced this world view with a new system of beliefs, Russian policy making was still, I would argue, heavily influenced by ideas and beliefs fostered and institutionalized in the Soviet period.

This becomes clearer when we zoom in on environmental policies. The Norwegian side, or more precisely the NME, adhered to the *principled belief* that protection of nature against industrial pollution is both morally and politically significant and that the 'polluter pays' principle applies. Its Russian counterparts in the environmental commission, and in the Pechenganikel negotiations, were still inclined to stress the primacy of industrial interests in present and future Russian society. Industry remained the principal building block for Russian prosperity, and this was true even after the Soviet epoch. Furthermore, they regularly insisted that the Norwegian approach to what became known as the 'death clouds' was hysterical. They themselves emphasized the continued existence of large tracts of untouched nature on the Kola Peninsula.

In their approach to the nickel industry, the Norwegians propagated the following *causal belief*: the pollution from Pechenganikel is a result of many years of inadequate Soviet maintenance and lack of technological development. This situation persisted, they would argue, because there were no environmental agencies that could enforce Soviet and Russian environmental regulations. Though the Russians were probably inclined to agree that technological advances had failed to materialize, their basic interpretation of the problem was very different: industrial pollution was seen in the Soviet way – as a symptom of a still incomplete production process, or suboptimal 'complex utilization' of the nickel ore. According to this rationale, the industry should not be subjected to strict laws that would threaten its existence: it should be left to develop and improve its technological solutions, and these would in the long run enable it to reach acceptable emissions levels.

Both the Norwegian and the Russian worldviews and their corresponding beliefs were long institutionalized in the respective political systems. While the Norwegian world view must be seen as still valid – that is, its basic tenets were still commonly adhered to – this was not the case in post-Soviet Russia. Despite belonging to an abandoned political system, the predominately Soviet ideas that still shaped Russian thinking about industrial pollution proved remarkably resistant. That no significant environmental measures to curtail emissions from Pechenganikel were taken, even after the industrial boom and improvement of living conditions in Russia after 2000, attests to this. Thus one can argue that it was the institutionalized ideas rather than the difficult economic situation of the 1990s that impeded a change in the Russian approach to industrial pollution.

Here, the dissonant Russian and Norwegian perceptions of each other and their conflicting approaches to environmental protection are for pedagogical purposes juxtaposed more sharply than they were probably experienced at the time. In actual negotiations, the chasm between the two parties was most likely less apparent. There is for example little to suggest that civil servants in the NME were fully aware of the general Russian distrust of Western motives as the 1990s moved on. Neither, I would argue, were the Russians aware of how widespread the Norwegian tendency to view them as underdeveloped foreign aid recipients who needed to be taught a lesson in environmental protection was. Especially when considering the discord in environmental questions, the Pechenganikel modernization can be understood as an attempt to reconcile two irreconcilable sets of ideas. This predicament was arguably deepened by the lack of understanding of the counterpart's beliefs and perceptions.

## Playing two-level games

The question remains, however, whether the *interests* that were at play in the modernization project could even out at least some of the differences in world views and beliefs. Could the two parties reach some common ground based on shared objectives, if not on a mutual understanding of the problem at hand? Let us return to Robert Putnam's model of *two-level games* that was briefly discussed in the introduction. To recap, Putnam's model stresses the influence exerted by domestic pressure groups on international negotiations. These groups, representing various interests that will either profit from or be harmed by certain outcomes of a negotiation, define what Putnam calls the *win-set*. The win-set is, in short, the range of possible negotiation outcomes acceptable to a sufficiently large number of domestic pressure groups. If such acceptance is pivotal to the conclusion of an agreement, the negotiating parties must have overlapping win-sets: for the negotiations to be successful, at least one outcome must have a chance of gaining approval in the respective domestic arenas.

To narrow in on the two win-sets that appeared when the Pechenganikel negotiations began, let us briefly review the interest groups on either side of the table. Starting with the Norwegian side, we find that this win-set was always strongly influenced by environmental objectives. The NOK 300 million support package was launched by the Norwegian government as an environmental initiative. The Norwegian bureaucracy, represented most notably by the NME, was adamant that Pechenganikel emissions be drastically cut, and was largely supported in this demand by the Norwegian population, by the domestic environmental movement and by local and central-level parliamentarians alike. In addition, there were the Norwegian industrial interests. Basically, they wanted a piece of the modernization cake, whether through large-scale contracts for the whole venture, or subcontracts for supplies. In this ambition, they were supported by local and central-level politicians and by the NME, the latter purportedly for environmental reasons. Thus, an agreement that would be accepted by the Norwegian pressure groups, or the Norwegian win-set, would look something like this: *Pechenganikel emissions will be cut by 90 per cent through a modernization led by a Norwegian entrepreneurial firm. Costs up to NOK 300 million will be paid by the Norwegian state. The rest will be covered by Russian sources.*

The Russian side of the equation comes out quite differently. While under negotiation, the Norwegian modernization initiative provoked various Russian concerns that were by and large incompatible with the Norwegian win-set. To a certain extent, the Russian environmental bureaucracy was sympathetic to the Norwegian demands. Other government agencies, however, were opposed to the implications of the proposed modernization. Most prominently, Russian customs and tax authorities were extremely reluctant to grant the necessary tax breaks to Norilsk Nikel/Pechenganikel. To the extent that the local population in Nikel was engaged, their main concern would be to preserve their workplaces, rather than improving the environment. Neither the top political level nor regional politicians were willing or able to fund their share of any of the modernization projects proposed during the 1990s, due not least to the fragile economic state of the Russian Federation throughout the decade. There was, however, considerable interest in the project among Norilsk Nikel and Pechenganikel managers. For industrial rather than environmental reasons, the concern and combine would be more than glad to see Norway support a major upgrading of the industrial installations in Pechenga. Thus, an agreement acceptable to the Russian side might look something like this: *Norway will fund the modernization of Pechenganikel. Small Russian contributions will come from the Norilsk Nikel concern or the Pechenganikel combine. Some taxes and customs will be lifted for the sake of the modernization, but the Russian state will not contribute direct funding.*

The distance between the two parties was immense. In effect, Norwegian authorities were demanding that, for the sake of good-neighbourly relations and with reference to principles of international environmental governance to which the Russians themselves only half-heartedly adhered, the Russians should spend immense amounts of money on a project of little or no interest to them.

To some extent, the stamina of Norwegian bureaucrats in this hopeless quest can be seen as a consequence of their playing a two-level game – they kept the project alive for the sake of domestic pressure groups that for various reasons had an interest in seeing the plans come to fruition. However, the main driver on the Norwegian side, I hold, was not the commercial or industrial interests involved, but a firm conviction of the rightness and fundamental moral rectitude of their demands. Add to this that the Norwegians met with some sympathy in Russia, and the strengthening of this conviction becomes apparent.

However, the groups that did support Norway's call for emissions reductions were in no position to exert decisive pressure. Russian environmental bureaucrats were, in terms of influence, dwarfed by colleagues in higher-status ministries. True, the nickel industry itself, motivated not by environmental concerns but by the need to upgrade Pechenganikel, kept backing the idea of a modernization scheme. Unaccompanied by any substantial financial commitment, though, this carried little weight in the bureaucratic corridors in Moscow. To most Russians who were involved, the crux of the matter was simple: the costly modernization of Pechenganikel was based almost exclusively on the interests and ideas of the small neighbouring state of Norway. There was, from the Russian perspective, no legal requirement or moral obligation that dictated them to spend large amounts of their scarce funds on this. Nor was there much to be won in terms of political benefit. In short, there was little for influential actors to gain by going along with the Norwegians, and correspondingly little to lose in hindering the modernization. Throughout the 1990s, the two win-sets never came close to reaching the overlap needed for a positive result.

After a decade of failed attempts at reaching common ground – or, put differently, at getting the Russian and Norwegian win-sets in the Pechenganikel modernization to overlap – the time was ripe for revamping the rules of the game. Two-level games are dynamic. Win-sets can become narrower or larger if, say, economic circumstances make certain negotiation outcomes less or more palatable. They can also be manipulated and thus enlarged. At the turn of the millennium, both things happened in the Pechenganikel modernization negotiations.

For one thing, the Russian economy was resurrected. The Russian boom had a positive impact on state coffers, and the transitional constraints of the 1990s were loosened. For Norilsk Nikel and Pechenganikel this coincided with a partial stabilization in the ownership struggle. To all intents and purposes, Pechenganikel found itself on a much firmer fiscal and managerial footing. Somewhat surprisingly, the company indicated a renewed interest in modernizing the installations, this time based on their own technology. Clearly, the improved economic circumstances had made the Russian win-set somewhat larger.

More important for the size of the win-sets on both sides was the manipulation of the two-level game that occurred at this point. Probably due largely to project fatigue, but also to provide a stringent monitoring system for project implementation, the NME outsourced, so to speak, the entire project to the NIB. Intentionally or not, the result was that several interest groups that had affected negotiations adversely were eliminated from the game. On the Russian side, there was from now on only one principal player, the Norilsk Nikel concern/ Pechenganikel combine. The Russian industrialists were motivated by the possibility of developing their own ore processing method. To the extent that technological improvements could lead them closer to a more 'complex utilization' of the nickel ore, their interests converged with the environmental objectives of the Norwegians.

At the other side of the negotiation table sat NIB, which had clear instructions from the NME but was nevertheless able to negotiate unencumbered by the political, environmental and industrial pressure groups that the state ministry routinely had to relate to. Thus, the Pechenganikel modernization became detached from the labyrinth of the two-level game and brought down to a few core issues. By leaving the negotiations and day-to-day management to NIB, the Norwegian government removed much of the impetus and opportunity for domestic actors to influence the project. No longer was the Pechenganikel modernization the object of negotiations between elected governments: now it was primarily a matter between a bank and an industrial actor. It was removed from the political sphere, thereby becoming less accessible to various interest groups. The Russian and Norwegian win-sets were enlarged and brought closer to overlap.

Initially, this removal of the Pechenganikel modernization from the world of politics seemed to work well. Under the new circumstances, the negotiations between NIB and Norilsk Nikel/ Pechenganikel reached a positive result approved by all parties. That said, the negotiations were not purely a matter between a bank and an industrial enterprise. The Norwegian and Russian governments had to agree on matters of both funding and taxation. As we saw in Chapter 5, reaching the necessary agreements here was easy – again, probably due to Russian economic revival after 2000.

Thus, harmony between Norwegian environmental concerns (drastic emissions cuts) and Russian industrial ambitions (development of domestic technology) had been established. The final solution to the environmental problems in the Pasvik valley seemed imminent. And yet, as we have seen, this was not to be. Simplifying the two-level game was not enough to save the modernization project. A seemingly straightforward path to implementation was obstructed by delays and Russian post-agreement bargaining. Failing to reach the interim project goals, the Russians now wanted to redirect the whole project in line with plans to relocate all smelting operations to Monchegorsk.

The Norwegians refused to contribute financially to such a solution. Whether this was a wise decision or not, is not for the present historical study to assess. What we can establish, though, is that the project fell through as a result of industrially motivated decisions (or deliberate lack thereof) within the Norilsk Nikel concern. The need to protect the natural surroundings of the Pechenganikel combine had no bearing on the management's deliberations on how to shape its future.

Rather, ecological issues were put firmly on the agenda only at the insistence of neighbouring countries. However, neither the Soviet Union nor its successor Russia nor the privatized Norilsk Nikel/ Pechenganikel would adopt environmental impulses from the outside world. Neither in the Soviet nor in the Russian contexts would the protection of nature be allowed to stand in the way of the economic and commercial necessities of production.

# Notes

## Chapter 1

1 Norwegian prime minister Jan Peder Syse made these remarks while speaking at an environmental rally organized by the 'Stop the Death Clouds from the Soviet Union'-campaign in Oslo Concert Hall on 10 September 1990. His address can be read in full in Christian Else and Henrik Syse (eds), *Ta ikke den ironiske tonen – tanker og taler av Jan P. Syse* (Oslo: Press forlag, 2003).

2 *Sør-Varanger Avis*, 12 February 2019.

3 Douglas R. Weiner, *A Little Corner of Freedom: Russian Nature Protection from Stalin to Gorbachëv* (Berkeley, CA: University of California Press, 1999), p. 444.

4 Jonathan D. Oldfield, *Russian Nature: Exploring the Environmental Consequences of Societal Change* (Aldershot: Ashgate, 2005), p. 130. Oldfield seems to overplay the significance of Western negative reports on Soviet ecology. To the extent that Russian scholars reached a Western audience, they too contributed to a confirmation of the bleak prospects. See for example Vladimir Kotov and Elena Nikitina, 'Russia in Transition: Obstacles to Environmental Protection', *Environment* 35, no. 10 (1993): 10–20.

5 Andy Bruno, *The Nature of Soviet Power: An Arctic Environmental History* (Cambridge: Cambridge University Press, 2016), pp. 267–75.

6 For a historical introduction to the emergence of international environmental politics, see Steinar Andresen, Elin Lerum Boasson and Geir Hønneland (eds), *International Environmental Agreements* (New York: Routledge, 2012), pp. 3–19.

7 Oldfield, *Russian Nature*, pp. 35–7.

8 Ibid., p. 67.

9 The idea of complex utilization of natural resources was also alive and well in the last years of the Soviet Union and in the early years of the Russian Federation (for an example from 2005, see Andy Bruno, *The Nature of Soviet Power* (Cambridge: Cambridge University Press, 2016),

p. 15 (footnote 22)). Three obvious examples can be found in the sources examined for this study. In 1986, when the Murmansk *oblast* authorities met with a delegation from the Department of Environmental Affairs at the Office of the Finnmark County Governor in Norway, the necessity of rational exploitation of natural resources was repeatedly stressed by Soviet delegates. For example, the head of the Murmansk Committee for Hydrometeorology and Environmental Monitoring (the closest to an environmental agency in Murmansk *oblast* at the time, subordinated to *Goskomgidromet*), Pyotr Vlasenko, underscored the need for 'rational management of the natural resources' and later stated that 'the work done by *[Goskomgidromet]* has contributed to a successful implementation in the USSR, thereunder in Murmansk oblast, of environmental measures aimed at improving the natural environment *while at the same time ensuring that its natural resources are maximally exploited in order to satisfy the needs of the Soviet population* [italics added]'. See Fylkesmannen i Finnmark (Office of the County Governor of Finnmark) (1987), 'Norsk/ Sovjetisk møte om miljøvern i felles grenseområder, Kirkenes 17.–19. juni 1986', Vadsø: Fylkesmannen i Finnmark, pp. 17 and 19. In 1988, the Soviet Union suggested that the environmental agreement with Norway contain the following sentence: 'This cooperation aims to solve important issues of nature protection and *problems concerning the efficient exploitation of natural resources* [italics added]'. The text was, at Norwegian insistence, changed to 'this cooperation aims to solve important issues of environmental protection and *to preserve the ecological balance* [italics added]'. See Geir Hønneland and Lars Rowe (2008), *Fra svarte skyer til helleristninger: Norsk–russisk miljøvernsamarbeid gjennom 20 år*, Trondheim: Tapir Akademisk Forlag, p. 29. In 1992, a Murmansk *oblast* planning document held that damages sustained by nature were to be understood as a result of the suboptimal complex utilization of minerals and bioresources. See Rune Castberg, 'Felles problem – ulik prioritering: nordisk–russisk miljøsamarbeid og nikkelverkene på Kola', *Nordisk Østforum* 2 (1993): 18. The Soviet and Russian understanding of industrial pollution will be discussed in further detail and juxtaposed to Norwegian views in Chapters 3 and 6.

10  According to Andy Bruno, Fersman firmly believed that man should conquer nature, endeavour to fully comprehend it through scientific

studies and even strive for a higher unity with it (See Bruno, 'Making Nature Modern: Economic Transformation and the Environment in the Soviet North', Ph.D. dissertation, University of Illinois: Urbana, IL, 2011, pp. 82–3. This thesis forms the basis for Bruno's book *The Nature of Soviet Power* from 2016, which is quoted above). The latter ambition seems oddly reminiscent of the ideas of post-war environmentalists (e.g. Aldo Leopold, *A Sand County Almanac* (Oxford: Oxford University Press, 1949)). However, as will be shown, Fersman's vision of nature was not one of a conservationist – quite the contrary.

11  Much of Fersman's significance as a prominent Soviet industrialist and geochemist is reflected in Bruno, 'Making Nature Modern'. For biographical details, see p. 35; for views on the human–nature relationship, see pp. 82–3; for 'complex utilization', see 109–11 and for impact on Soviet environmental tradition, see p. 280.

12  Weiner, *A Little Corner of Freedom*, p. 450.

13  Judith Goldstein and Robert O. Keohane, 'Ideas and Foreign Policy: An Analytical Framework', in *Ideas and Foreign Policy: Beliefs, Institutions and Political Change*, ed. J. Goldstein and R. O. Keohane (Ithaca, NY: Cornell University Press, 1993), pp. 3–7. Goldstein and Keohane polemicize against political scientists, of whom they quote many, for emphasizing only interests in their analyses of international relations.

14  Ibid., p. 5.

15  Ibid., p. 8.

16  It is impossible to do justice to Goldstein and Keohane's model in a brief presentation like this, which is only meant to highlight some elements of specific relevance to the present study. For a full elaboration of the theory, with illustrative empirical studies, see Judith Goldstein and Robert O. Keohane (eds), *Ideas and Foreign Policy: Beliefs, Institutions and Political Change* (Ithaca: Cornell University Press, 1993). The above presentation is based on Goldstein and Keohane's introductory chapter 'Ideas and Foreign Policy: An Analytical Framework' (pp. 3–30). The quoted passage is taken from page 21.

17  See Robert D. Putnam, 'The Logic of Two-Level Games', *International Organization* 42, no. 3 (1988): 427–60. In fact, a game-theoretical analysis of the Pechenganikel negotiations has already been published. That work is, however, driven almost exclusively by theory, is not very accessible

and has scant relevance for the present study. Here I am referring to Margrethe Aanesen, 'To Russia with Love? Four Essays on Public Intervention under Asymmetric Information: The Petsjenganikel Case on the Kola Peninsula', Ph.D. dissertation, Norut report 08/2006, Tromsø: Norut Samfunnsforskning AS, 2006.

18  On win-sets, see Putnam, 'The Logic of Two-Level Games', p. 440.

# Chapter 2

1  Stortingsproposisjon nr. 1 (2008–9), Budsjetterminen 2009, kap. 1400–72.

2  Stortingsproposisjon nr. 1 (2010–11), Budsjettterminen 2011, kap. 12, resultatområde 5, målområde 3: nord- og polarområdene.

3  As quoted in Leonid A. Potemkin, *U severnoi granitsy: Pechenga sovetskaya* (Murmansk: Murmanskoe Knizhnoe izdatelstvo, 1965), p. 235. This is only one of many examples of the lofty and enthusiastic rhetoric used about the settling of Zapolyarnyi. Potemkin's highly patriotic book is flush with quotes from individual youth communists praising their new life in the North. Incidentally, there were not only *komsomoltsy* in Zapolyarnyi. The work force, especially the initial settlers, had been ordered from other areas, and some young men, after having completed their military service in the North, worked for a period in Zapolyarnyi before they could return to their homes. See Potemkin, *U severnoi granitsy*, pp. 228–33.

4  Ibid., p. 227.

5  Nikel was in Soviet terminology defined as a 'city-like village' (*poselok gorodskogo tipa*).

6  Potemkin, *U severnoi granitsy*, p. 248.

7  M. B. Smirnov (ed.), *Sistema ispravitelno-trudovykh lagerei v SSSR* (Moscow: Zvenya, 1998), pp. 432–3. This reference book about the system of Soviet correctional labour camps was published by the organization Memorial and the State Archive of the Russian Federation. The entry referred to statements that the railroad was completed by the fifth construction battalion, although it remains unclear whether this unit was part of the

Gulag system or another administration. However, it seems likely (as briefly mentioned in Chapter 5) that Pechenga's proximity to the national border would temper the use of prison labour, as the possible escape of the prisoners must have been a concern.

8  For a thorough discussion of this process, see Lars Rowe, *Industry, War and Stalin's Battle for Resources: The Arctic and the Environment* (London and New York: I.B. Tauris 2020), pp. 67–103.

9  V. A. Matsak (ed.), *Pechenga. Opyt kraevedcheskoi entsiklopedii* (Murmansk: Prosvetitelskiy tsentr 'Dobrokhot', 2005), p. 424.

10  See Heikki Väyrynen, *Petrologie des Nickelerzfeldes kaulatunturi-kammikivitunturi in Petsamo* (Helsinki: Commission Geologique de Finlande, 1938), Bulletin No. 116.

11  O. Ya. Galushko, '40 let trudovoy vakhty gorno-metallurgicheskogo kombinata "Pechenganikel"', *Tsvetnaya Metallurgiya* no. 9 (1985): 1–3.

12  See Rowe, *Industry, War and Stalin's Battle for Resources*, pp. 109–12.

13  Holtsmark (ed.), *Norge og Sovjetunionen* (1995), p. 365 (doc. 281). The Soviet intention was to build a power plant over the Kolttaköngäs rapids (Norwegian: Skoltefoss) in the Borisoglebskii area.

14  Hanne Brusletto, 'Forhandlinger mellom Norge og Sovjetunionen om kraftutbygging i Pasvikelven 1945–1963: Norsk–sovjetisk brobygging under den kalde krigen', MA thesis in history, University of Oslo, fall 1994, pp. 17–38. The Soviet Union, incidentally, never intended to supply the Norwegian side with hydropower. See AVPRF, f. 0116, op. 28, p. 130, d. 19, l. 17, letter from Ramzaitsev in the Ministry of Foreign Trade to MID and Glavnikelkobalt, 14 May 1946.

15  See Rowe, *Industry, War and Stalin's Battle for Resources*, pp. 109–12.

16  Due to unstable water levels in the Pasvik River, which flooded occasionally, Norwegian agricultural areas and livestock in the Pasvik valley had suffered. The shifting water levels were attributed to the construction of the Jäniskoski and Rajakoski power plants. Norway therefore initiated trilateral talks with Finland and the Soviet Union to establish a new regulation regime. See Holtsmark, *Norge og Sovjetunionen*, documents 341, 345, 369 and 370; Brusletto, 'Forhandlinger mellom Norge og Sovjetunionen', pp. 41–3.

17  Brusletto, 'Forhandlinger mellom Norge og Sovjetunionen'.

18  Brusletto, 'Forhandlinger mellom Norge og Sovjetunionen', pp. 116–36.

19  In fact, the first initiative to open up the Borisoglebskii enclave came from Norwegian actors engaged in the front organization Norwegian–Soviet friendship society, aiming to promote relations between Norway and the Soviet Union – or rather, propagate Soviet worldviews to the Norwegian public. During its short spell as a tourist destination, the Borisoglebskii enclave became mostly known for cheap access to liquor and tobacco, which quickly gave the local bar a bad reputation. See Ingunn Rotihaug, '"For fred og vennskap mellom folkene": Sambandet Norge–Sovjetunionen 1945–70', in *Defence Studies 1/2000* (Oslo: Norwegian Institute for Defence Studies, 2000), pp. 79–85.

20  Soviet depictions of socialism and communism as the only truly peace-promoting societal system were used in propaganda already in the 1930s and even more fervently during the final years of Stalin's life. See Marshall D. Shulman, *Stalin's Foreign Policy Reappraised* (Cambridge, MA: Harvard University Press, 1963). For an analysis of the Soviet peace offensive in Norway, see Lars Rowe, '"Nyttige idioter"? Fredsfronten i Norge, 1949–1956', *Defence Studies 1/2002* (Oslo: Norwegian Institute for Defence Studies, 2002).

21  For a discussion of Soviet motives as seen from the Norwegian side, see Brusletto, 'Forhandlinger mellom Norge og Sovjetunionen'.

22  See AVPRF, f. 0116, op. 53, p. 200, d. 13, ll. 10–17, letter from Soviet ambassador to Norway Lunkov to Kovalev in MID, 11 January 1963. It is remarkable how the official Soviet views were incorporated in the quasi-scientific writing on the issue as well. An example is Potemkin's *U severnoi granitsy*, which is in all respects in line with the official Soviet interpretations of reality.

23  Potemkin, *U severnoi granitsy*, p. 269.

24  See Rowe, *Industry, War and Stalin's Battle for Resources*, pp. 96–101.

25  O. Ya. Galushko, '40 let trudovoy vakhty gorno-metallurgicheskogo kombinata "Pechenganikel"', pp. 1–3.

26  Kristian Gerner, 'Naturmiljö och politik i Österuropa', *Nordisk Østforum* 2 (1993): 5–14. For a thorough discussion of various explanations of the origins of environmental degradation, see Bruno, 'Making Nature Modern', pp. 191–201.

27  A pioneer study exposing Soviet environmental problems was Marshall I. Goldman's book *The Spoils of Progress*, published in 1972. Goldman

opened up new perspectives on the socialist handling, or lack thereof, of industrial waste. He also demonstrated how the notion that privately owned industry was at the core of the problem of pollution was not a new one. Already in 1912, the English economist A. C. Pigou had presented his theory about the lack of incentives for pollution control in a capitalist economy. For a presentation of Pigou's thinking, see Marshall I. Goldman, *The Spoils of Progress: Environmental Pollution in the Soviet Union* (Cambridge, MA: MIT Press, 1972), pp. 18ff.

28   See for example Goldman, *The Spoils of Progress*, pp. 22–3. Also in Gro Elisabeth Olsen, 'Moja po tvoja? Vilkår for samarbeid med Russland: en studie av Petsjenganikel-prosjektet 1985–2002', MA thesis in history, University of Bergen, spring 2005, p. 21, a brief review of literature on the topic is given.

29   Boris Komarov, *The Destruction of Nature in the Soviet Union* (White Plains, NY: M. E. Sharpe, 1980), pp. 62–74. This book comes across as somewhat combative and overly negative of the Soviet system. Nevertheless, it offers interesting insights into the malfunctions of Soviet legislation and lists various examples of how nature suffered from intensive industrialization. It was later translated into several languages.

30   Grigorii Svirskii, "Komandirovka v Nikel', *Ogonek* no. 24 (1960): 20–1.

31   Kristian Gerner and Lars Lundgren, *Planhushållning och miljöproblem: Sovjetisk debatt om natur och samhälle 1960–1976* (Stockholm: LiberFörlag, 1978), p. 151.

32   Several reasons are listed for the rather surprising choice of transporting ore from Siberia (Taimyr Peninsula) to the Kola Peninsula – a transport that required the costly creation and operation of a fleet of ice breakers to keep the Northern Sea Route open. For a thorough discussion, see Andrew R. Bond and Richard M. Levine, 'Air Pollution Problem at Monchegorsk Continues', *Soviet Geography* 30, no. 3 (1989): 255–61; and Andrew R. Bond, 'The Russian Copper Industry and the Noril'sk Joint-Stock Company in the Mid-1990s', *Post-Soviet Geography and Economics* 37, no. 5 (1996): 286–329.

33   Document from NME, on file with the author.

34   Castberg, 'Felles problem – ulik prioritering', pp. 15–24.

35   See Steinar Andresen, Elin Lerum Boasson and Geir Hønneland, 'Framveksten av internasjonal miljøpolitikk', in *Internasjonal miljøpolitikk*,

ed. Steinar Andresen, Elin Lerum Boasson and Geir Hønneland (Bergen: Fagbokforlaget, 2008), pp. 17–36, for presentations of the various 'green waves' in international relations.

36 The foremost expression of the rapprochement between the United States and the Soviet Union during détente was the Strategic Arms Limitation Talks (SALT I and II). Commencing in the late 1960s, and progressing through the 1970s, these talks, and the ensuing agreements, were designed to reduce the overall number of nuclear warheads. Brezhnev's main motivation for these talks, however, was probably not disarmament itself, but rather the need to unburden Soviet economy. For a brief review of Soviet domestic motivations for détente, see Richard Sakwa, *Soviet Politics: an Introduction* (London: Routledge, 1989), pp. 86–8.

37 Vladimir Kotov and Elena Nikitina, 'Implementation and Effectiveness of the Acid Rain Regime in Russia', in *The Implementation and Effectiveness of International Environmental Commitments: Theory and Practice*, ed. David G. Victor, Kai Raustiala and Eugene B. Skolnikoff (Cambridge, MA: MIT Press, 1998a), pp. 519–20; Robert G. Darst, *Smokestack Diplomacy: Cooperation and Conflict in East–West Environmental Politics* (Cambridge, MA: MIT Press, 2001), pp. 94–8.

38 Darst, *Smokestack Diplomacy*, p. 23. The Soviet reluctance to make any concessions on human rights issues, even if this was a clear provision in the Helsinki Accords, is vividly illustrated by the case of Soviet Jewry and their fight to exercise fundamental human rights, such as freedom to emigrate. See Gal Beckerman, *When They Come for Us, We'll Be Gone: The Epic Struggle to Save Soviet Jewry* (Boston, MA: Houghton Mifflin Harcourt, 2010).

39 In hindsight, this visit and the resulting cooperation between Norwegian and Soviet bureaucrats have been described as an important impetus to the subsequent establishment of the Soviet–Norwegian environmental commission ten years later. Interview with Valentin Sokolovskii, Moscow, 10 April 2007. Sokolovskii was deputy chairman in the Soviet State Committee for Hydrometeorolgy (Goskomgidromet) and later the State Committee for Environmental Protection (Goskompriroda). He also served as head of the Soviet delegation to the joint Soviet–Norwegian Environmental Commission from 1988 to 1991.

40 Kotov and Nikitina, 'Implementation and Effectiveness of the Acid Rain Regime in Russia', pp. 520–1; Darst, *Smokestack Diplomacy*, pp. 96–102.

41 Soviet commitment to the LRTAP regime did not, of course, offset the negative consequences of other and more prominent foreign policy problems at the time, the most important being Western protests at the continued Soviet war in Afghanistan after the invasion in December 1979.

42 Geir Hønneland and Anne-Kristin Jørgensen, *Implementing International Environmental Agreements in Russia* (Manchester: Manchester University Press, 2003), pp. 159–60. This book makes a clear theoretical distinction between 'compliance' and 'implementation'. Thus, the authors conclude, commenting on the specific case of the Pechenganikel combine and with reference to the lack of legal measures against its sulphurous emissions, that 'Russia can show a high degree of formal compliance with its LRTAP commitments in the 1990s, but a correspondingly poor implementation record in the same period' (p. 160).

43 Darst, *Smokestack Diplomacy*, pp. 100–6.

44 Kotov and Nikitina, 'Implementation and Effectiveness of the Acid Rain Regime in Russia', p. 519. The authors refer to various academic works where similar findings are presented.

45 Darst, *Smokestack Diplomacy*, p. 25.

46 See Elana Wilson Rowe, 'Who Is to Blame? Agency, Causality, Responsibility and the Role of Experts in Russian Framings of Global Climate Change', *Europe-Asia Studies* 61, no. 4 (2009): 593–619, where several proponents of this interpretation of Russia's climate politics are cited.

47 See Stephen Kotkin, *Armageddon Averted: The Soviet Collapse 1970–2000* (Oxford: Oxford University Press, 2001), pp. 67–8.

48 Weiner, *A Little Corner of Freedom*, p. 1ff.

49 Douglas Weiner has shown that in some instances the Soviet conservationist movement did in fact confront industrial interests. See Douglas Weiner, *Models of Nature: Conservation and Community Ecology in the Soviet Union, 1917–1935* (Bloomington, IN: Indiana University Press, 1988).

50 The Soviet version of civil society (*obshchestvennost*) was, due to the inner logic of the one-party state, quite different from the reality experienced by Western NGOs. In practice, these organizations were kept well within an acceptable range of what Soviet authorities would tolerate. Thus, the organizations became mostly a quasi-critical element dependent on official approval. In more recent times this phenomenon, which still

exists, has been dubbed GONGOs, or governmental non-governmental organizations. For a discussion of Soviet and post-Soviet civil society, see Vadim Volkov, "'Obshchestvennost'": Russia's Lost Concept of a Civil Society', in *Civil Society in the Baltic Sea Region*, ed. Norbert Götz and Jörg Hackmann (Aldershot: Ashgate, 2003), pp. 63–72.

51  Weiner, *A Little Corner of Freedom*, p. 437.

52  Darst, *Smokestack Diplomacy*, p. 110. The environmental activist organization Bellona more than implies that there is a direct link between the nickel industry and the degree of political activism in its company towns: 'Unfortunately, in cities where Norilsk Nickel plants are located, the environmental and human rights movement is weak.' See Larisa Bronder et al., 'Norilsk Nickel: The Soviet Legacy of Industrial Pollution', Bellona report 2010, Oslo: Bellona Foundation, 2010, p. 5.

53  See for example Kristian Åtland, 'Russisk nordområdepolitikk etter den kalde krigen: Forholdet mellom næringsinteresser og militærstrategiske interesser', *FFI-report 2003/00713*, Kjeller (FFI), 2003, pp. 7–8; and Olsen, 'Moja po tvoja?', pp. 31ff.

54  Mikhail S. Gorbachev, *Izbrannye rechi i stati*, vol. 5 (Moscow: Izdatelstvo politicheskoi literatury, 1988), pp. 337–8.

55  Fylkesmannen i Finnmark, *Norsk/Sovjetisk møte om miljøvern i felles grenseområder, Kirkenes 17.–19. juni 1986* (Vadsø: Fylkesmannen i Finnmark, 1987).

56  Geir Hønneland and Lars Rowe, *Fra svarte skyer til helleristninger: Norsk-russisk miljøvernsamarbeid gjennom 20 år* (Trondheim: Tapir Akademisk Forlag, 2008), p. 27.

57  Darst, *Smokestack Diplomacy*, pp. 106–7. See also MD 1997/2749, m. 3, unnumb. doc, telefax from ET to NME, dated 13 January 1992.

58  MD 1997/2749, m. 5, competition announcement from Minpriroda RSFSR, dated 2 June 1993.

# Chapter 3

1  According to Jonathan D. Oldfield, much of the Western academic representation of Soviet and later Russian (post-Soviet) environmental problems has been fraught with generalizations, and thus created a

one-dimensional and exaggerated picture of a natural environment completely bereft of protection. While Oldfield acknowledges that the 'Soviet system bequeathed extensive environmental problems with global ramifications', he aims in his study *Russian Nature* to differentiate the established image of the Soviet and post-Soviet area as an environmental disaster zone in lack of nature management. See Jonathan D. Oldfield, *Russian Nature: Exploring the Environmental Consequences of Societal Change* (Aldershot: Ashgate, 2005), pp. 1ff. I would agree with Oldfield concerning the need for this. Nevertheless, the present study focuses on one specific environmental 'disaster zone' and will therefore inevitably contribute to the aforementioned one-dimensionality rather than to a differentiation of the picture.

2 For a discussion of the first period of the environmental commission, see Hønneland and Rowe, *Fra svarte skyer til helleristninger*, pp. 28ff.

3 Fylkesmannen i Finnmark, 'Norsk/Sovjetisk møte', pp. 21–37 (Per-Einar Fiskebeck's lecture).

4 Fylkesmannen i Finnmark, 'Norsk/Sovjetisk møte', p. 6.

5 Geir Hønneland, *Russia and the West: Environmental Co-operation and Conflict* (London: Routledge, 2003), p. 114.

6 Goskomgidromet represented the Soviet Union in the Cooperative Programme for Monitoring and Evaluation of Long-Range Transmission of Air Pollutants in Europe (EMEP), which was part of the LRTAP regime (see Chapter 2). Through the operation of a meteorological monitoring centre, Goskomgidromet contributed to international data exchange on aerial transportation of pollutants. See Kotov and Nikitina, 'Implementation and Effectiveness of the Acid Rain Regime in Russia', p. 524. Goskomgidromet (formerly Gidromet) was originally a state meteorological service (*sluzhba*), but was elevated to a state committee in 1978, and at the same time given the added responsibility for environmental monitoring. See Hønneland, *Russia and the West*, p. 153 (endnote 2).

7 For a brief introduction to the emergence of the Norwegian environmental bureaucracy, see Andresen, Lerum Boasson and Hønneland, 'Framveksten av internasjonal miljøpolitikk', pp. 30–4.

8 Andresen, Lerum Boasson and Hønneland, *International Environmental Agreements*, pp. 12–13.

9  While Brundtland's environmental track record may have its flaws in the eyes of Norwegian environmental activists, there is little doubt that she strongly contributed to a general acceptance in Norwegian politics of the relevance of environmental concerns. For the emergence of a Norwegian environmental policy and Brundtland's role in this, see Andresen, Lerum Boasson and Hønneland, 'Framveksten av internasjonal miljøpolitikk', pp. 30–4.

10  As quoted in Castberg, 'Felles problem – ulik prioritering', p. 18. Castberg here quotes a draft for the Murmansk oblast social and economic programme, authored well into the post-Soviet era in 1992. There are many examples of this type of descriptions of environmental problems in Soviet documents. The chosen quote from 1992, however, illustrates what is a main point in the following discussion: that traditional Soviet perceptions of nature persisted in the glasnost era and beyond. The notion that pollution in fact stemmed from suboptimal exploitation of natural resources derived from the model for maximal utilization of minerals developed by Kola Peninsula industrial pioneer Aleksandr Fersman. See introduction and Bruno, 'Making Nature Modern', pp. 109–11, on how Soviet perceptions of nature persisted in the glasnost era and beyond.

11  Philip R. Pryde, *Environmental Management in the Soviet Union* (Cambridge: Cambridge University Press, 1991), p. 12.

12  Hønneland and Rowe, *Fra svarte skyer til helleristninger*, p. 29.

13  For the minutes of the meeting, see Fylkesmannen i Finnmark, 'Norsk/Sovjetisk møte om miljøvern'.

14  MD 1987/3497, doc. 1, letter from the NME to the Norwegian Ministry of Foreign Affairs, 3 July 1987, with attachment: revised draft for environmental agreement between Norway and the USSR (in Norwegian translation), received in Moscow at the beginning of June 1987.

15  For the final text of the agreement in Norwegian, see Hønneland and Rowe, *Fra svarte skyer til helleristninger*, pp. 172–6.

16  For a discussion of Soviet perceptions of nature, see Oldfield, *Russian Nature*, pp. 30–3.

17  This heading is taken from colleague Geir Hønneland's insightful discourse analysis of perceptions and behaviour in the environmental interface between post-Soviet Russia and Norway. See Hønneland, *Russia and the West*, pp. 117–19.

18 Ibid., p. 118.

19 The following description of Nikel is the author's own, based on several visits there, and meant to give the reader a visual impression of the place, to allow a better understanding of press reports. The author makes no claims to deep aesthetic or architectural understanding but tries for a factual description of Nikel and surroundings and the impressions they made on visiting Westerners in the late 1980s and early 1990s. For a photographic presentation of Nikel, see Ola Solvang, *Under de tre pipene: Fotografier fra Nikel* (Tromsø: Bladet Nordlys, 1998).

20 Valentin Sokolovskii, who was to head the Soviet delegation to the Soviet–Norwegian environmental commission, would express strong irritation with the number of Norwegian journalists who participated on the trip, claiming that 'about 90% of the minister's entourage was made up of journalists'. See MD 1989/6084, doc. 33, memo from conversation with Sokolovskii at Norwegian embassy in Moscow, dated 16 July 1991.

21 *Verdens Gang*, 1 July 1988.

22 *Aftenposten*, 19 August 1988.

23 *Nordlys*, 19 August 1988.

24 The tendency to 'oversee' elements, like pristine nature, that might weaken the depiction of the Kola Peninsula as an ecological disaster zone, has been durable. Pål Skedsmo makes a parallel observation as late as 2004. See Pål Skedsmo, *Russisk sivilsamfunn og norske hjelpere* (Trondheim: Tapir Akademisk Forlag, 2010), pp. 36–7.

25 Minutes from the first meeting in the joint Soviet–Norwegian Environmental Commission, 23–26 August 1988. On file with the author.

26 *Aftenposten*, 27 August 1988; *Nordlys*, 27 August 1988.

27 Hønneland, *Russia and the West*, pp. 119–21. Hønneland refers here to what he calls the 'anti-hysteria discourse', which is juxtaposed with the 'death clouds discourse'.

28 *Aftenposten*, 27 and 29 August 1988; *Nordlys*, 27 and 29 August 1988.

29 *Sør-Varanger Avis*, 13 May 1989.

30 Interestingly, the substantial local pollution stemming from AS Sydvaranger had never raised local emotions to a significant level. The local community was possibly too dependent on the company for emploment to raise the issue. On the few occasions when voices were raised against the cornerstone enterprise, these were efficiently muted by the local political leadership. See Arvid Ellingsen, 'Aksjon "Stopp

dødsskyene fra Sovjet" – en umulig lederoppgave', MA thesis, University of Bergen, 1997, pp. 38–9.

31 These four key figures were Kåre Tannvik, Thorbjørn Bjørklie, Hans M. Møllersen and Tor Aarnes. The following account of the action group is based on Ellingsen, 'Aksjon "Stopp dødsskyene fra Sovjet"', pp. 44ff.

32 As quoted in ibid., p. 47.

33 Ibid., p. 19.

34 Hønneland, *Russia and the West*, p. 119.

35 Not only the press and activists, but also local authorities in Sør-Varanger, employed the 'death clouds discourse'. An example can be found in MD1989/6084, doc. 7, statement from Sør-Varanger municipal council 11 June 1990, where it is said that there is a 'grave danger that Sør-Varanger in time will become an inhospitable desert as the areas around Nikel and Zapolyarnyi are today'.

36 MD 1987/3497, doc. 14, telefax from the Norwegian Embassy in Moscow to the NME, dated 21 August 1989, with attachments; attached minutes from informal conversation after meeting.

37 See Bond, 'The Russian Copper Industry', p. 306. Bond's figures are based on Russian sources. Norwegian figures are higher. For example, a report from the Norwegian pollution control agency SFT authored in 2001 states that the Pechenganikel emissions in 1990, a year previous, were 250,000 tons (MD 2000/3598, m. 3, d. 40, 'Assessment of PERG's report on modernization of Pechenganikel', p. 4). Nevertheless, Bond's Russian figures, although perhaps flawed, do illustrate the main point in this context: that Pechenganikel was considered a minor polluter when compared to Norilsk Nikel.

38 The campaign 'Stop the Death Clouds from the Soviet Union' did collaborate with a group in Pechenga called EKOS. See Ellingsen, 'Aksjon "Stopp dødsskyene fra Sovjet"'. EKOS seems to have quickly faded away, however. Pechenganikel director Blatov later described the local environmentalism in Nikel as initially strong, but quickly fading. Referring to the high (in a Russian context) wages received by Pechenganikel workers, he said that Nikel citizens were more worried about losing their well-paid jobs than the environmental degradation caused by the smelting plant. See MD 1997/2749, m. 8, memo from the Norwegian Embassy to Russia, minutes from conversation with

Director Baltov and technical director Gulevich, dated 10 August 1995.
For a discussion of environmental activism on the Kola Peninsula and
Norwegian support to NGOs there (and some of the ulterior motives that
seem to have spurred this activism), see Skedsmo, *Russisk sivilsamfunn
og norske hjelpere.*

39  Local reliance on Pechenganikel for income was overwhelming. Being
the only viable source of livelihood in the area, the combine provided
about 80 per cent of Pechenga *raion*'s tax revenues. Also, the Murmansk
*oblast* administration was heavily dependent on revenues from the nickel
industry, with 40 per cent of the budget covered by taxes from nickel-
related business. See Vesa Rautio, 'Petsamo – "Kaipaukseni maasta"
globaalitalouden pyörteisiin', *Terra* 112, no. 3 (2000): 129–40.

40  Darst, *Smokestack Diplomacy*, p. 36; Andresen, Lerum Boasson and
Hønneland, *International Environmental Agreements*, p. 12.

41  Darst, *Smokestack Diplomacy*, p. 36.

42  Finnish–Soviet relations after the Second World War were remarkable
in many ways. Although in a precipitous position as a small state along
the Soviet border, Finland was never incorporated into the Eastern Bloc,
but remained a parliamentary democracy. On the other hand, Finnish
sensitivity to Soviet security concerns was readily demonstrated in
the country's foreign policy, which was conducted within the limiting
framework of the Soviet–Finnish Agreement of Friendship, Cooperation
and Mutual Assistance. Thus, a string of compromises regulated the
relationship between a survival-minded Finnish state and a Soviet
Union wary of the small nation's well-documented resistance to pressure
from the outside. For a Finnish perspective on what has been called,
misleadingly according to the author, 'Finlandization', see introduction in
Max Jacobson, *Finland Survived: An Account of the Finnish–Soviet Winter
War 1939–1940* (Helsinki: Otava, 1984). For a glimpse into the Soviet
approach to Finland in the first post-war years, see Vladislav Zubok
and Constantine Pleshakov, *Inside the Kremlin's Cold War: From Stalin
to Khrushchev* (Cambridge, MA: Harvard University Press, 1996), pp.
116–19.

43  Until the end of the 1980s, the Soviet Union alone accounted for almost
20 per cent of Finnish exports. Thus, when the Soviet economy crashed
in the late 1980s, this had a deep impact on the Finnish economy.

In combination with a series of other unfortunate factors, the Soviet downturn led to a severe economic crisis in Finland in the early 1990s. See Jukka Nevakivi, 'From the Continuation War to the Present 1944–1999', in *From Grand Duchy to a Modern State: A Political History of Finland since 1809*, ed. Osmo Jussila, Seppo Hentilä and Jukka Nevakivi (London: Hurst & Company, 1999), pp. 331–2.

44 Castberg, 'Felles problem – ulik prioritering', pp. 18–20; Olsen, 'Moja po tvoja?', pp. 40–1.

45 MD 1997/2749, m. 1, memo from the office of Norway's Prime Minister (hereafter SMK, from the Norwegian designation *Statsministerens kontor*) to NME, 4 September 1990; MD 1997/2749, m. 4, document titled 'Pechenganikel smelter project' issued by Outokumpu Technology OY, August 1992, p. 1.

46 Castberg, 'Felles problem – ulik prioritering', pp. 16–17.

47 Protocol from first meeting of the Soviet–Norwegian Environmental Commission, 23–26 August 1988. On file with the author.

48 MD 1987/3497, doc. 6, letter from the Office of the County Governor in Finnmark to NME, dated 8 March 1989.

49 Protocol from the second meeting in the Soviet–Norwegian Environmental Commission, 10–14 April 1989. On file with the author.

50 MD 1997/2749, m. 1, NME memo on talking points for conversation with Soviet minister of environment, dated 11 May 1990, with attached memo on measures against pollution from the Kola Peninsula.

51 MD 1987/3497, documents 8, 12 and 13, various correspondence between NME, the Office of the County Governor of Finnmark and the Norwegian companies regarding the delegation exchange, May and August 1989.

52 MD 1987/3497, doc. 14, telefax from the Norwegian Embassy in Moscow to NME, dated 21 August 1989, with attachments; attached minutes from informal conversation after meeting. The Norwegian government was already in the process of introducing environmental subsidies as a measure against transboundary pollution damaging Norwegian territory. The debate in the Norwegian parliament in the fall 1989 would result in a substantial effort both in direct payments and loan guarantees to clean up industrial sites in Poland. See Ellingsen, 'Aksjon "Stopp dødsskyene fra Sovjet"', pp. 30–1.

53 Pål Kolstø, *Kjempen vakler: Sovjetunionen under Gorbatsjov* (Oslo: Universitetsforlaget, 1990), pp. 269–70.

54 Ibid., pp. 271–2.

55 Basically, Norilsk Nikel replaced Mintsvetmet as the controlling holder of the vertically organized complex comprising all enterprises involved in the production of Soviet nickel. These were Norilsk Nikel in Norilsk (mining and metallurgy), Pechenganikel in Pechenga (mining and metallurgy), Severonikel in Monchegorsk (metallurgy), the Krasnoyarsk Nonferrous Metals Plant (alloys and precious metals), the Olenegorsk Mechanical Plant (machine-building and equipment repair) and the Gipronikel Planning and Design Institute in St. Petersburg. In addition, the state concern gained control over a comprehensive infrastructure supporting the nickel industry. For further details, see Bond, 'The Russian Copper Industry', pp. 295–6.

56 Kolstø, *Kjempen vakler*, pp. 294–5.

57 There was reciprocity in this matter: Norwegians visiting the Soviet Union (and later Russia – the reciprocity lasted well into the 1990s) received a lump sum in roubles upon arrival. These roubles did not bear the same significance to the Norwegians as the Norwegian currency did to Soviet (and later Russian) delegates. Rather, the arrangement was symbolically important to uphold the notion that the cooperation was like-sided and not simply foreign aid. That personal pecuniary gain was a central motivational factor for Soviet delegates in this period is, due to its sensitive character, difficult to establish. However, there is massive Soviet/Russian as well as Norwegian anecdotal evidence. Furthermore, the author's own experience within the field strongly supports the assumption. An early example of *per diem* allowances paid out to Soviet delegates while in Norway can be found in MD 1987/3497, doc. 7, letter from the Office of the County Governor of Finnmark to NME, dated 28 April 1989 and doc. 10, letter from the Office of the County Governor of Finnmark to NME, dated 21 June 1989, with attachment reading 'Received pocket money 6 days á NOK 150,– Total NOK 900,– Kirkenes 11.6.1989' with the signature of three Soviet delegates.

58 This view is supported by other researchers. See for example Castberg, 'Felles problem – ulik prioritering'. The Finnish commercial motivation has, somewhat speculatively, been seen in connection with a subsequent

Finnish attempt to lease the Pechenga area and Eastern Karelia. See Olsen, 'Moja po tvoja?', pp. 45–6.

59 Syse, Syse and Syse, *Ta ikke den ironiske tonen*, pp. 171–3.

60 For the sake of simplicity, sums will be referred to in US dollars in the following, although Norwegian kroner, Finnish marks, Soviet rubles and US dollars are all used in the source material. The currencies are converted according to historical exchange rates on the Oanda website (www.oanda.com). For the conversion of sums from late 1989, exchange rates from January 1990 are used, as these are the earliest available.

61 MD 1997/2749, m. 1, memo (AIP), dated 17 September 1990 with attached memo from Outokumpu titled 'The renovation proposition for Kola nickel smelters by Outokumpu', dated 14 September 1990.

62 MD 1197/2749, m. 1, memo (SMK) about Nordic meeting concerning the Kola initiative, 4 October 1990, dated October 1990, p. 9.

63 Ibid., memo (AIP), 'follow-up to the Kola initiative', dated 7 September 1990. See also interview with Outokumpu's managing director Voutilainen in the Finnish newspaper *Kauppalehti*, 15 October 1990. Here Voutilainen dismisses any possibility of increased efficiency or bettered quality of end product as a result of the project, which he describes as fundamentally environmental. The interview can be found in Swedish translation as attachment to MD 1997/2749, m. 1, dispatch from Norwegian embassy in Helsinki, dated 19 October 1990.

64 Ibid., memo (SMK) dated 4 October 1990 about Nordic meeting concerning the Kola initiative 4 October 1990, p. 11. The size of direct support from the two governments was not set in stone, but would depend on various factors, most importantly the Russian choice of technology.

65 See for example MD 1997/2749, m. 1, memo (AIP), dated 22 June 1990 and ibid., memo (Norwegian Ministry of Finance), dated 7 August 1990, where the OECD agreement is discussed. The Soviet Union was still at this time in OECD terminology defined within Category I, or in other words an industrialized country, as opposed to a developing country, and was therefore not entitled to aid or subsidies. See MD 1997/2749, m. 1, memo (SMK), dated 7 August 1990.

66 See for example MD 1997/2749, m. 1, dispatch (SMK) to MD (AIP), dated 6 September 1990. In fact, the Finnish and Norwegian direct

contributions were in effect intended as an interest rate subsidy. Once this money was transferred to a fund under NIB's control, the recipient would be free to use them, according to NME, to 'bring the interest load down to a level that is acceptable to Soviet authorities'. See MD 1997/2749, m. 2, memo (AI)), dated 7 March 1991.

67  Ibid., m. 1, memo (AIP), dated 17 September 1990.

68  Ibid., memo (AIP), 'Status for the Kola project – questions to be answered', dated 5 November 1990.

69  Ibid., letter from GIEK to NME, dated 15 November 1990.

70  Ibid., memo (AIP), 'Status for the Kola project – questions to be answered', dated 5 November 1990.

71  Ibid., letter from ET to NME, dated 23 January 1991.

72  Ibid., memo (AIP/SMK), dated 4 October 1990. In the interview with Outokumpu's managing director Voutilainen in the Finnish newspaper *Kauppalehti*, 15 October 1990 (available in Swedish translation in MD 1997/2749, m. 1, attached to memo (AIP), 'follow-up to the Kola initiative', dated 7 September 1990), the foremost domestic consumers of nickel are said to be the steel industry and the aviation industry, both significant to the Soviet armed forces.

73  Interview with Outokumpu's managing director Voutilainen in *Kauppalehti*, 15 October 1990, found in Swedish translation in MD 1997/2749, m. 1, attached to memo (AIP), 'follow-up to the Kola initiative', dated 7 September 1990.

74  MD 1997/2749, m. 1, memo (AIP), dated 27 June 1990.

75  Ibid. Elkem Technology, owned by the Elkem concern, a metallurgical company involved in production of aluminium and ferroalloys, was set up to provide Elkem itself and external customers with environmental process technology. See MD 1997/2749, m. 1, memo (ET), June 1990.

76  The Vanyukov furnace, named for metallurgical specialist Vladimir A. Vanyukov (1880–1957), had been installed in the Norilsk smelting plants in the mid-1980s. See company journal *Norilskii Nikel*, vol. 35, no. 4, June/July 2007.

77  MD 1997/2749, m. 1, memo (ET), June 1990.

78  Ibid., memo (MFA), dated 23 June 1990.

79  See *Metal Bulletin*, 21 June 1990, p. 7, where an Elkem source, interviewed about the company's talks with the Soviet nickel industry,

is quoted as saying that 'environmentally-friendly smelting operations were a "huge market" now that more firms were jumping on the ecology bandwagon and Elkem would seem to be well-placed to take advantage of this trend'.

80 MD 1997/2749, m. 1, letter of intent attached to memo (ET), 'Status for Kola project', dated 29 August 1990.

81 *Hufvudstadsbladet*, 26 September 1990.

82 MD 1997/2749, m. 1, letter from Syse to prime ministers Harri Holkeri (Finland), Nikolai Ryzhkov (the Soviet Union) and Ingmar Carlsson (Sweden), dated 24 August 1990.

83 Ibid., memo (SMK), dated 31 August 1990.

84 Ibid., memo (SMK), dated 3 September 1990.

85 Ibid., memo (AIP), dated 7 September 1990.

86 MD 1997/2749, m. 2, letter from Sør-Varanger municipality to Prime Minister Gro Harlem Brundtland, dated 11 December 1990; MD 1997/2749, m. 2, letter from Sør-Varanger Faglige Samorganisasjon to Prime Minister Gro Harlem Brundtland, dated 21 May 1991; ibid., letter from Norske Entreprenørers service-organisasjon to NME, dated 18 June 1991.

87 MD 1997/2749, m. 2, memo (AIP) from meeting between state secretaries, dated 5 December 1990.

88 A brief but insightful account of the events of 1990–1991 that led to the collapse of the USSR can be found in Robert Service, *A History of Modern Russia from Nicholas II to Vladimir Putin* (Cambridge, MA: Harvard University Press, 2005), pp. 488–508. On the emergence of Russian statehood, see Richard Sakwa, *Russian Politics and Society*, 3rd edn (London: Routledge, 2002), pp. 16–19.

89 MD 1997/2749, m. 1, memo (AIP), dated 7 September 1990. See also Sakwa, *Russian Politics and Society*, p. 18.

90 MD 1997/2749, m. 1, memo (AIP), dated 6 September 1990.

91 MD 06, unnumbered doc., minutes from conversation between Igor Gavrilov and Ambassador Dagfinn Stenseth, dated 20 September 1990; MD 1989/6084, doc. 16, minutes from conversation between Yuri Fokin and Ambassador Dagfinn Stenseth, dated 12 September 1990. Interestingly, Fokin was later to become the ambassador of the Russian Federation to Norway.

92  MD 1997/2749, m. 2, memo (AIP), 8 January 1991.

93  Ibid., memo (AIP), dated 12 March 1991.

94  Ibid., memo (AIP), dated 16 January 1991.

95  Ibid., memo (AIP), dated 12 March 1991.

96  MD 06, unnumbered doc., minutes from conversation between heads of oblast administrations in Murmansk, Arkhangelsk and Vologda and Prime Minister of the Republic of Karelia and Ambassador Dagfinn Stenseth, 13 March 1991, dated 18 March 1991; MD 1997/2749, m. 2, report from Norwegian embassy envoy's journey to Murmansk, 11–14 March 1991, dated 19 March 1991.

97  *Aftenposten*, 17 August 1991.

98  MD 1989/6084, doc. 33, minutes from conversation between Valentin Sokolovskii and Ambassador Dagfinn Stenseth, dated 16 July 1991.

99  The Soviet–Norwegian Environmental Commission had a consultative meeting in Svanvik in the Norwegian Pasvik valley, in late August 1991. Modernization of the nickel industry was one of the foremost questions discussed. This took place at the same time as the situation in Moscow was growing extremely tense, with the coup attempt against Gorbachev. See Hønneland and Rowe, *Fra svarte skyer til helleristninger*, pp. 46–7.

100 MD 1989/6084, doc. 36, letter from NME to the Norwegian embassy to USSR, dated 26 September 1991.

101 MD 1997/2749, m. 2, letter from ET to NME, dated 7 October 1991.

102 MD 1997/2749, m. 3, memo (AIP), dated 14 November 1991.

103 The most comprehensive plans originated in Sør-Varanger municipality, where the local mechanical shipyard Kimek organized a host of local enterprises and enterprises from elsewhere in Norway to support the modernization project with a wide range of services. See MD 1997/2749, m. 2, letter w/attachments from Kimek to NME, dated 4 October 1991.

104 MD 1997/2749, m. 3, telefax from Finnish Ministry of Foreign Affairs to NME, dated 8 October 1991.

105 MD 1989/6084, doc. 39, memo (AIP), dated 29 October 1991.

106 MD 1989/6084, doc. 40, memo (AIP), dated 25 October1991.

107 In light of the lack of progress in the modernization project, several Norwegian interest groups publicly pressed for higher intensity in the proceedings. See for example MD 1997/2749, m. 2, letter from Svanvik Folkehøgskole to Norwegian prime minister, dated 6 March 1991; ibid.,

m.3, telefax from 'Stop the Death Clouds from the Soviet Union', dated 6 November 1991; ibid., declaration from women's group in Troms Labor Party, dated 24 November 1991. In addition, 'Stop the Death Clouds from the Soviet Union' continued to exert pressure through various media channels.

108 MD 1989/6084, unnumbered doc., memo (SMK), dated 29 October 1991.

109 There was concern that the imminent collapse of the Soviet Union meant that the Environmental Commission would fall apart. See *Aftenposten*, 6 November 1991; *Nordlys*, 7 November 1991.

110 MD 1989/6084, doc. 47, memo (AIP), dated 8 November 1991.

111 MD 1989/6084, doc. 48, memo (AIP), dated 8 November 1991; MD 1997/2749, m. 3, memo (AIP), dated 14 November 1991; *Aftenposten*, 8 November 1991.

112 *Aftenposten*, 8 November 1991.

# Chapter 4

1 For an introduction to this interpretation, see Lars Rowe and Geir Hønneland, 'Norge og Russland: Tilbake til normaltilstanden', *Nordisk Østforum* 24, no. 2 (2010): 133–47.

2 For a discussion of this phenomenon – 'the power of the gift' – within the context of Norwegian support to environmental organizations in the post-Soviet area, see Skedsmo, *Russisk sivilsamfunn og norske hjelpere*.

3 For an excellent overview of the history of Norwegian development aid, see Helge Ø. Pharo. 'Reluctance, Enthusiasm and Indulgence: The Expansion of Bilateral Norwegian Aid', in *The Aid Rush: Aid Regimes in Northern Europe during the Cold War*, ed. Helge Ø. Pharo and Monika Pohle Fraser (Oslo: Unipub, 2008), pp. 53–89. I have, together with colleagues at the Fridtjof Nansen Institute, discussed the imagined inverted asymmetry and the Norwegian foreign aid discourse in several publications. For the environmental collaboration, see Lars Rowe, Geir Hønneland and Arild Moe, 'Evaluering av miljøvernsamarbeidet mellom Norge og Russland', *FNI report* 7/2007. Lysaker: Fridtjof Nansen Institute, 2007, pp. 4–7; for the Norwegian–Russian health collaboration,

see Geir Hønneland and Lars Rowe, *Health as International Politics: Combating Communicable Diseases in the Baltic Sea Region* (Burlington, VT: Ashgate, 2004); and for a more general debate, see Geir Hønneland and Lars Rowe, *Nordområdene – hva nå?* (Trondheim: Tapir Akademisk Forlag, 2010), pp. 45–6. For an example of overblown Norwegian ambitions towards Russia, see Utenriksdepartementet, 'Om handlingsprogrammet for Øst-Europa', Oslo: Utenriksdepartementet, 1994–1995, p. 53, where it is stated that Norway's overriding goal with the support to the 'reform countries' (Eastern Europe, including Russia) was to 'contribute to a fundamental restructuring of these societies with the aim of securing a democratic and economically sustainable development'.

4 The Soviet–Norwegian environmental commission was re-established as part of the Russian–Norwegian environmental agreement, replacing the Soviet–Norwegian one, signed 3 September 1992. Environmental talks, however, had been going on at uneven intervals throughout the transitional period. Norway was the first country to recognize the Russian Federation as an independent nation state on 16 December 1991. See Hønneland and Rowe, *Fra svarte skyer til helleristninger*, pp. 53–6.

5 MD 1997/2749, m. 3, unnumbered doc., fax from Norwegian Embassy in Helsinki to NME, dated 8 January 1992.

6 MD 1989/6048, unnumbered doc., memo (AIP), dated 21 January 1992; MD 1997/2749, m. 3, memo (ET), dated 28 January 1992; ibid., memo (AIP), dated 23 January 1992.

7 MD 1997/2749, m. 3, unnumbered doc., fax from ET to NME, dated 13 January 1992.

8 Ibid., memo (ET), dated 28 January 1992.

9 See Olsen, 'Moja po tvoja?', pp. 55–6.

10 MD 1989/6084, unnumbered doc., memo (AIP), dated 21 January 1992; ibid., doc. 55, memo (AIP), dated 29 January 1992; ibid., doc. 56, memo (AIP), dated 21 February 1992.

11 MD 1997/2749, m. 4, memo (AIP), dated 19 February 1992.

12 MD 1997/2749, minutes from meeting in the Russian Ministry of Economic Affairs, dated 5 June 1992.

13 MD 1997/2749, m. 4, document titled 'Pechenganikel smelter project', issued by Outokumpu Technology OY, dated August 1992, p. 1.

14  MD 1997/2749, minutes from meeting in the Russian Ministry of Economic Affairs, dated 5 June 1992.

15  MD 1997/2749, m. 4, memo (AIP), dated 11 September 1992.

16  Ibid., memo (AIP), dated 23 November 1992.

17  Ibid. The environmental ministers met in Copenhagen in November 1992 to discuss the big issue in environmental politics at the time – the thinning ozone layer.

18  MD 1997/2749, m. 4, minutes from the meeting of Ministers of Environment in Copenhagen 24 November 1992, p. 1.

19  MD 1997/2749, m. 5, memo (AIP), dated 14 December 1992; ibid., memo (AIP), dated 25 February 1993.

20  MD 1997/2749, m. 6, memo (AIP), dated 10 January 1994.

21  MD1997/2749, m. 6, letter from NN to NME, dated 3 February 1994.

22  MD 1997/2749, m. 6, memo (AIP), dated 24 March 1994.

23  MD 1997/2749, m. 6, memo (AIP), dated 1 June 1994. Interestingly, though, one of the participant companies in PRC, Kvaerner Engineering, owned several large Finnish subsidiaries that were potential subcontractors. This, one may surmise, was one of the reasons that Kvaerner Engineering was included in PRC. See MD 1997/2749, m. 6, letter from Boliden Contech to the Swedish Ministry for the Environment and Natural Resources, dated 11 January 1994.

24  MD 1997/2749, m. 6, letters from Thorbjørn Berntsen to Finnish minister of the environment Sirpa Pietikäinen and Swedish minister of the environment Olof Johansson, both dated 8 June 1994.

25  See Margrethe Aanesen, 'To Russia With Love? Four Essays on Public Intervention under Asymmetric Information: The Petsjenganikel case on the Kola Peninsula', Ph.D. dissertation, Norut report 08/2006, Tromsø: Norut Samfunnsforskning AS, 2006, p. 33.

26  MD 1997/2749, m. 6, fax from ET to NME, dated 1 September 1994.

27  Ibid.; also fax from Boliden Contech to NME, dated 12 September 1994.

28  MD 1997/2749, m. 6, letter from Boliden Contech to the Swedish Ministry for the Environment and Natural Resources, dated 11 January 1994; ibid., fax from ET to NME, dated 22 February 1994.

29  MD 1997/2749, m. 5, memo (AIP), not dated, but content places it in late September/early October 1994. See also ibid., m. 6, Contract between NME and Elkem Technology, dated 25 October 1994.

30  MD 1997/2749, m. 6, fax from ET to NME, dated 22 February 1994.

31 MD 1997/2749, m. 6, memo (AIP), dated 5 December 1994.

32 MD 1997/2749, m. 6, memo (GIEK), dated 5 January 1995.

33 MD 1997/2749, m. 7, file with all minutes from Thorbjørn Berntsen's Moscow trip January 23–26, 1995, introductory memo (AIP); ibid., memo (AIP), minutes from meeting with Norilsk Nikel management January 24, 1995.

34 MD 1997/2749, m. 7, file with all minutes from Thorbjørn Berntsen's Moscow trip January 23–26, 1995, memo (the Norwegian Embassy to Russia), dated 1 February 1995.

35 According to Norilsk Nikel management, the Russian state's role in the development of the nickel industry has been very unclear, and therefore counterproductive. See Rautio, 'Petsamo – "Kaipaukseni maasta" globaalitalouden pyörteisiin'.

36 Sakwa, *Russian Politics and Society*, pp. 287–91.

37 MD 1997/2749, m. 7, file with all minutes from Thorbjørn Berntsen's Moscow trip January 23–26 1995, various memos.

38 MD 1997/2749, m. 7, fax from the Norwegian Embassy to Russia to NME, dated 9 February 1995; ibid., fax from Norwegian Embassy to Russia to NME, dated 15 February 1995.

39 MD 1997/2749, m. 7, memo (AIP), minutes from meeting with Blatov, dated 6 March 1995.

40 Sakwa, Russian Politics and Society, p. 83.

41 MD 1997/2749, m. 7, memo (SMK), dated 14 March 1995.

42 MD 1997/2749, m. 7, fax from the Norwegian Embassy to Russia to NME, dated 23 March 1995; ibid., letter from Norilsk Nikel to PRC, dated 15 March 1995.

43 I have not seen the letter, but its content and date of dispatch (15 March 1995) is known from other sources. See MD 1997/2749, m. 7, memo (Norwegian MFA, copied to a number of recipients), dated 23 April 1995; ibid., fax from the Norwegian Embassy to Russia, dated 23 March 1995.

44 MD 1997/2749, m. 7, memo (Norwegian Embassy to Russia), dated 11 May 1995; ibid., memo (Norwegian Embassy to Russia), dated 16 May 1995.

45 *Rossiiskaya Gazeta*, 13 July 1995. The decree was dated 1 July 1995 and signed by Prime Minister Viktor Chernomyrdin himself, which was a crucial expression of its importance. Interestingly, the Norwegians do

not seem to have received information about the decree until it was published in the Russian government newspaper.

46 MD 1997/2749, m. 8, memo from the Norwegian Embassy to Russia, minutes from meeting with Blatov and Gulevich, dated 10 August 1995.

47 MD 1997/2749, m. 7, minutes from meeting between NME, Kværner Engineering and Boliden Contech, dated 13 July 1995.

48 MD 1997/2749, m. 8, memo (AIP), dated 26 July 1995.

49 MD 1997/2749, m. 8, memo from the Norwegian minister of foreign affairs to the Norwegian prime minister and minister of the environment, dated 21 July 1995.

50 MD 1997/2749, m. 7, memo (AIP), dated 29 July 1995; ibid., memo (AIP), dated 24 July 1995.

51 MD 1997/2749, m. 8, fax from Elkem (ET's mother company) to NME, dated 21 August 1995; ibid., memo (AIP), dated 26 July 1995, minutes from meeting between ET and Thorbjørn Berntsen.

52 In the sources, there is also mention of possible rivalry between Elkem Technology and Kværner Engineering having contributed to Elkem Technology's refusal to renew the consortium agreement. See for example MD 1997/2749, m. 8, memo (AIP), dated 26 July 1995, minutes from meeting between ET and Thorbjørn Berntsen. This factor is never fully expanded on, however.

53 MD 1997/2749, m. 8, memo from Norwegian embassy to Russia, minutes from meeting with Gulevich and Blatov, dated 10 August 1995.

54 MD 1997/2749, m. 8, memo (SMK), dated 25 July 1995.

55 MD 1997/2749, m. 7, memo (Norwegian embassy to Russia), dated 11 May 1995.

56 MD 1997/2749, m. 7, memo (SMK), dated 11 May 1995.

57 MD 1997/2749, m. 8, memo (AIP), dated 26 July 1995.

58 Ibid. The meeting in London concerned the situation in Bosnia.

59 MD 1997/2749, m. 7, memo (Norwegian embassy to Russia), dated 20 July 1995.

60 MD 1997/2749, m. 8, memo from the Ministry of Foreign Affairs, dated 8 August 1995.

61 MD 1997/2749, m. 8, memo (AIP), dated 11 August 1995.

62 MD 1997/2749, m. 8, memo (the Ministry of Foreign Affairs), minutes from conversation with Ambassador Fokin, dated 25 August 1995.

For a Norwegian version of the memorandum text, see ibid., attachment
to memo (the Ministry of Foreign Affairs), dated 30 August 1995.

63  MD 1997/2749, m. 8, memo (PRC), dated 23 February 1996.

64  MD 1997/2749, m. 8, letter from Kværner Engineering to Norilsk Nikel,
dated 18 August 1995.

65  MD 1997/2749, m. 8, memo (AIP), dated 17 August 1995.

66  MD 1997/2749, m. 8, memo (AIP), dated 29 September 1995.

67  MD 1997/2749, m. 8, memo (AIP), dated 20 October 1995.

68  MD 1997/2749, m. 8, letter from Kværner Engineering (CEO) to Norilsk
Nikel (chairman of the board), dated 18 October 1995; ibid., memo
(AIP), points for letter to President Yeltsin, dated 20 October 1995; ibid.,
memo (AIP), minutes from meeting between Thorbjørn Berntsen and
Erik Tønseth (CEO Kværner) dated 24 October 1995; ibid., letter from
NME to Minister Kossov in the Russian Ministry for the Economy, dated
2 November 1995.

69  MD 1997/2749, m. 8, memo (AIP), dated 13 November 1995.

70  MD 1997/2749, m. 8, memo (PRC), dated 1 November 1995.

71  One illustrative example: when Norilsk Nikel/Pechenganikel asked PRC
to develop a 'basic engineering study' that would take into account
their new plans for processing Norilsk ore, the study was meant
to inform the technical planning for more sulphur in the process.
Once the study had been completed, however, Pechenganikel proved
unable to pay for the service. As the PRC project leader from Kværner
Engineering laconically commented: 'The project is outside K[værner]
E[ngineering]'s regular business area and has a customer that represents
a business culture that is different from what K[værner] E[ngineering]
is accustomed to.' See MD 1997/2749, m. 8, memo (PRC), dated 23
February 1996.

72  MD 1997/2749, m. 8, memo (PRC), dated 23 February 1996.

73  MD 1997/2749, m. 8, memo (AIP), dated 21 March 1996.

74  MD 1997/2749, m. 8, letter from KE to NME, dated 2 April 1996.

75  *Aftenposten*, 23 March 1996.

76  MD 1997/2749, m. 8, memo (AIP), dated 10 April 1996. An unsigned
copy of the agreement (protocol) is attached to the memo.

77  Norsk Telegrambyrå, 25 March 1996.

78  *Verdens Gang*, 27 March 1996. In February 1996, a group of Russian
moguls who controlled most of the Russian media joined forces to

help Yeltsin to a second term. For a discussion of their role in Yeltsin's eventual victory, see Sakwa, *Russian Politics and Society*, pp. 83ff.

79 *Dagens Næringsliv*, 27 March 1996.

80 MD 1997/2749, m. 9, memo (AIP), dated 23 May 1996. Blatov's undated letter is attached to this memo.

81 MD 1997/2749, m. 9, memo (AIP), dated 23 May 1996. Berntsen's letter to Danilov-Danilyan, dated 24 May 1996, is attached to this memo.

82 MD 1997/2749, m. 9, letter from Danilov-Danilyan to Berntsen, dated 29 May 1996; ibid., memo (AIP), dated 24 June 1996.

83 Ibid., memo (AIP), dated 3 July 1996.

84 Ibid., memo (Norwegian Ministry of Foreign Affairs), dated 5 July 1996.

85 Ibid., memo (Norwegian Embassy to Russia), dated 10 July 1996.

86 Ibid., memo (KE), dated 18 July 1996.

87 Ibid., memo (minutes from meeting between KE and NME), dated 4 September 1996.

88 Ibid., memo (AIP), dated 23 October 1996; ibid., protocol from Government Conference proceedings, dated 31 October 1996.

89 Ibid., m. 9, letter from KE to MD, dated 30 October 1996.

90 MD 1997/2749, m. 9, Protocol from government conference proceedings, 21–22 November 1996.

91 Ibid., memo (AIP), dated 24 February 1997.

92 Ibid., letter from KE to NME, dated 12 December 1996; ibid., letter from NME to KE, dated 17 January 1997; ibid., letter from KE to NME, dated 28 January 1997.

93 MD 1997/2749, m. 9, memo (Norwegian Embassy to Russia), dated 18 February 1997.

94 Ibid., memo (AIP), dated 24 February 1997.

95 Ibid., fax from Khloponin to NME, KE, BC and ET, dated 17 April 1997.

96 Ibid., m. 9, fax from Khloponin to NME, KE, BC and ET, dated 17 April 1997.

97 Ibid., m. 9, memo (NIB), dated 5 June 1997.

98 Ibid., m. 9, memo (AIP), dated 12 August 1997.

99 Ibid., m. 10, memo (AIP), dated 4 September 1997. The process of privatization of the Norilsk Nikel state concern is discussed in Chapter 5.

100 For an overview of the Norilsk Nikel concern's various plans for development of its industrial activity in the late 1990s and early 2000s,

see Andrew R. Bond and Richard M. Levine, 'Noril'sk Nickel and Russian Platinum-Group Metals Production', *Post-Soviet Geography and Economics* 42, no. 2 (2001): 77–104.

101 MD 1997/2749, m. 10, memo (Storvik & Co. AS), dated 11 August 1997; ibid., memo (Storvik & Co. AS), dated 14 December 1997.

102 *Pechenga*, 24 December 1997. The newspaper article, which gives a fairly detailed description of Pechenganikel's development potential at the end of 1997, although without touching on the environmental problems, is found in English translation in MD 1997/2749, m. 10.

103 MD 1997/2749, m. 10, press release from Outokumpu, dated 18 December 1997. See also newspaper article in *Pechenga*, 24 December 1997 (see previous footnote) for additional information about the Pechenga–Outokumpu agreement.

104 The extent of remaining ore resources in Pechenga was an issue fraught with uncertainty. In 1996, the NME was informed by the Geological Survey of Norway (NGU) that Pechenganikel only had sufficient deposits for another five years of production. See MD 1997/2749, m. 9, letter from NGU to NME, dated 21 August 1996. This, of course, was in contrast to the company's own projections. See for example newspaper article in *Pechenga*, 24 December 1997. The assessments from mid-1990s to late 1990s were even further off the appraisal in 2010, when the NME received information from Norilsk Nikel indicating that ore reserves were sufficient for another sixty years of production (see Chapter 5).

105 See MD 1997/2749, m. 10, memo (AIP), dated 22 January 1998. Many other examples could be added.

106 MD 1997/2749, m. 10, assessment report from the Norwegian Pollution Control Authority (Statens forurensingtilsyn), dated 11 August 1998.

107 Anders Åslund, 'Russia's Financial Crisis: Causes and Possible Remedies', *Post-Soviet Geography and Economics* 39, no. 6 (1998): 309–28; Mario Blejer, 'Financial Crisis in Russia: A Comment', *Post-Soviet Geography and Economics* 39, no. 6 (1998): 329–31; Anders Åslund, 'Russia's Economic Transformation under Putin', *Eurasian Geography and Economics* 45, no. 6 (2004): 397–420.

108 Åslund, 'Russia's Economic Transformation under Putin'.

109 Åslund, 'Russia's Financial Crisis'.

110 Olsen, 'Moja po tvoja?', p. 88.

111  MD 1997/2749, m. 10, memo (the Norwegian Consulate in Murmansk), dated 4 November 1998.

112  For an interesting discussion of how the state–company relationship developed, see Olsen, 'Moja po tvoja?', pp. 70–2.

113  MD 1997/2749, m. 10, memo (the Norwegian Consulate in Murmansk), dated 4 November 1998.

114  MD 1997/2749, m. 10, memo (AIP), dated 15 January 1999.

115  MD 1997/2749, m. 10, memo (PR section), minutes from meeting between NME, NIB and the Norwegian Pollution Control Authority (SFT) 18 February 2000.

116  MD 1997/2749, m. 10, memo (AIP), dated 2 June 2000; ibid., memo (NIB), dated 21 June 2000.

117  The post-Soviet rhetoric contained many references to how Russia, the other Soviet republics and the Eastern European states were freed of the communist burden and thus enabled to reunite with the Western world. An additional element in the Norwegian discourse on post-Soviet Russia was what Geir Hønneland has called the 'dramaturgy of reunification'. Citing historical bonds with Norwegian settlers on the Kola Peninsula and age-old trade relations (the pomor trade) between the White Sea area and Eastern Finnmark, Norwegian (and some Russian) actors claimed that the time had finally come to 'pick up where we left off in 1917'and resume close contacts between 'the northerners' in Russia and Norway. See Geir Hønneland, *Barentsbrytninger: Norsk nordområdepolitikk etter den kalde krigen* (Kristiansand: Høyskoleforlaget, 2005), pp. 107–8. For a more detailed discussion of the conceptualization and critique of this 'dramaturgy of reunification' and the idea of a northern commonality between Russians and Norwegians, see Geir Hønneland, *Borderland Russians: Identity, Narrative and International Relations* (Basingstoke: Palgrave Macmillan, 2010).

118  The establishment of BEAR was a typical example of the region-building strategy in vogue in international politics in the North in the first half of 1990s. Other regional cooperative bodies created in this period were the Council of the Baltic Sea States (1992) and the Arctic Council (1996). A later initiative came in 1997, when Finland launched the 'Northern Dimension' within the European Union. For an analysis of this trend, see Iver B. Neumann, 'A Region-Building Approach to Northern Europe', *Review of International Studies* 20, no. 1 (1994): 53–74.

119 For a discussion of the BEAR cooperation from a Norwegian perspective, see Hønneland, *Barentsbrytninger*.

120 Ibid., p. 44.

121 MD 1997/2749, m. 4, memo (AIP), dated 16 September 1992.

122 MD 1997/2749, m. 5, memo (AIP), dated 14 December 1992.

123 For example, in MD 1997/2749, m. 5, memo (AIP), dated 25 February 1993; ibid., memo (AIP), dated 3 March 1993; ibid., m. 6, memo (AIP), dated 12 December 1994; ibid., m. 8, memo (AIP), dated 29 September 1995.

124 See Hønneland, *Barentsbrytninger*, pp. 123–49.

125 The privatization of Norilsk Nikel has been meticulously, if quite emotionally, described over more than a thousand pages by Aleksandr Korostelev in his book from 2008: *Delo 'Norilskii Nikel': Privat-kapitalizm Rossi. Strategiya, taktika i metody promyshlennoi privatizatsii po Chubais. Priroda proiskhozhdeniya oligarkhicheskikh kapitalov* (Moscow: Algoritm-Kniga). A scholarly discussion of the role of the oligarchs is provided in Stephen Fortescue, *Russia's Oil Barons and Metal Magnates: Oligarchs and the State in Transition* (Basingstoke: Palgrave Macmillan, 2006). Neither author leaves any doubt about the legally dubious nature of the privatization processes.

126 The shares-for-credit scheme was developed to allow the bankrupt Russian state to borrow money from private financial institutions against security in state shares in some resource-based industries, Norilsk Nikel among them. When the loans were defaulted, the private banks were free to auction the shares, and in the case of Norilsk Nikel Oneksimbank made sure to win the bid. See Fortescue, *Russia's Oil Barons and Metal Magnates*, pp. 28–29 (on Potanin's background) and pp. 54ff (on shares-for-credit).

127 Oleg Soskovets, who acted as deputy prime minister from 1993, appointed Vladimir Potanin to the post of deputy minister of industry in 1996. See Olsen, 'Moja po tvoja?', pp. 72–4.

128 See for example Olsen, 'Moja po tvoja?', p. 101.

129 Bruno, *Making Nature Modern*, pp. 240–1. While Bruno does not claim that privatized Norilsk Nikel can be considered environmentally 'good' by any standard, he does make a case for how the concern has improved environmental standards after privatization, that is, become 'better'.

130 For an interesting analysis of inertia in post-Soviet mining management, see Veikko Kärnä, 'A Return to the Past? An Institutional Analysis of

Transitional Development in the Russian Mining Industry', *Series A-5:* 2007, Turku: Publications of the Turku School of Economics, 2007.

131 Fortescue, *Russia's Oil Barons and Metal Magnates*, pp. 86–7.

132 The concept gained wide acclaim and was used by other scholars and journalists alike. On the Russian and Western media application of the phrase, see Stephen F. Cohen, *Failed Crusade: America and the Tragedy of Post-Communist Russia* (New York: Norton, 2000), p. 104.

133 Through numerous interviews with Russian participants in various fields of the Russian–Norwegian cooperation, the author and colleagues at the Fridtjof Nansen Institute have repeatedly encountered Russian suspicions of foul play in connection with Norwegian project funding. Though varying in intensity, this distrust played an increasingly important role in the bilateral relationship as the 1990s drew to a close. For an example from the Russian–Norwegian health collaboration that covers Russian scepticism to Norwegian objectives, see Geir Hønneland and Lars Rowe, 'Western versus Post-Soviet Medicine: Fighting Tuberculosis and HIV in North-West Russia and the Baltic States', *Journal of Communist Studies and Transition Politics* 21, no. 3 (2005): 395–414. For a comprehensive discussion of this development in the most central areas of Russian–Norwegian cooperation, see Hønneland and Rowe, *Nordområdene – hva nå?*.

# Chapter 5

1 For an interesting discussion of the fiscal turnaround during Putin's first years in power, see Åslund, 'Russia's Economic Transformation under Putin', and Marshall I. Goldman's comments to Åslund's article (Goldman, 'Anders in Wonderland: Comments on Russia's Transformation under Putin', *Eurasian Geography and Economics* 45, no. 6 (2004): 429–34). The main quarrel is whether it was Putin's policies as such or, as Goldman maintains, primarily soaring oil prices that made Russia's economic recovery possible. Whatever the eventual outcome of this debate, which will be further illuminated by future historical studies, there is little doubt that Putin's personal popularity was immense throughout the 2010s. In this period, his political stature allowed him to present himself to the Russian electorate not only as the champion of Russia's comeback in

international politics, but also as the main architect behind the country's economic revival.

2 The literature on Vladimir Putin's presidency is far too extensive to be discussed here. Furthermore, the present study entertains no ambitions to illuminate this specific part of the recent Russian past. The assessments that do come to the fore, especially the claim that Putin's governance style draws heavily on both pre-revolutionary and Soviet traditions, are based in the author's own studies as reflected in chapter 2 in Hønneland and Rowe, *Nordområdene – hva nå?*, pp. 35–55.

3 This paragraph is based on MD 1997/2749, m. 10, letter (the Norwegian Consulate in Murmansk), dated 24 July 2000.

4 Stortingsproposisjon nr. 1 (1991–2), Budsjetterminen 1992 (the Norwegian National Budget for 1992), chapters 1400–71.

5 MD 1997/2749, m. 10, memo (NIB), dated 21 June 2000.

6 The proposal for reallocation of the money was based on the misunderstanding that the Syse government had earmarked 300 million NOK for the purpose of the Pechenganikel modernization. This funding existed only on paper, as a promise in the national budget that presupposed that an environmentally satisfactory solution was presented, and thus was never 'set aside', so to speak, in a designated fund.

7 MD 1997/2749, m. 10, memo (AIP), dated 29 June 2000.

8 MD 1997/2749, m. 10, press release from Norilsk Nikel, dated 4 August 2000.

9 See MD 1997/2749, m. 10, dispatch to which press release was attached, dated 8 August 2000. That Norwegian authorities were unaware of the announcement's content before it was released becomes apparent from the dispatch, where it is stated that 'the [Norwegian] embassy has contacted the company [Norilsk Nikel] and asked for further information about the plan'. When it comes to Norwegian assessment of the Vanyukov technology, this was to happen in the fall and winter 2000/2001, as we shall see in the following.

10 MD 1997/2749, m. 10, letter from SFT to NME (AIP), dated 16 August 2000 with two attached reports from SFT, one assessing various environmental efforts on the Kola Peninsula and the other weighing the environmental potential of alternative spending of the Norwegian support funds.

11 Norilsk Nikel's insistence on a high degree of secrecy is mentioned many times in the documentation, and principles of confidentiality for the assessment group were laid down in PERG's terms of reference. See MD 2000/2627-5, letter from NIB to NME, dated 5 September 2000. The ToR are attached to the letter. The Vanyukov technology, which had originally been developed from the 1950s onwards, was at the time being installed at several of Norilsk Nikel's industrial sites. However, it had not yet been used in nickel refining, only for copper production. See MD 2000/3598, m. 1, minutes from meeting held 10 November 2000 between NIB and Norwegian ministries, dated 22 November 2000.

12 MD 2000/2627-5, letter from NIB to NME, dated 5 September 2000.

13 MD 2000/2627-11, PERG's draft report, dated 31 January 2001 (Norwegian version).

14 In the sources, the sums of both USD 30 million and 270 million NOK are used in referring to the Norwegian contribution at this time. However, the Norwegian government had already sponsored earlier efforts, most notably PRC projecting, by 30 million NOK, and was only prepared to contribute the remainder of its original support sum of 300 million NOK. Whether this sum equalled USD 30 million or not depended on the exchange rate at any given time. At the end of 2002, when the new project was in its very start, the exchange rate (1USD=6.9277NOK) in fact meant that the Norwegian support was almost USD 39 million. See MD 2000/3598, m. 2, Pechenganikel Modernisation Program: Report to donors, dated 28 February 2003.

15 The first mention of the trust fund idea in the documents dates from early October 2000. See MD 2000/2627, minutes from meeting between NIB and NME, dated 6 October 2000.

16 This type of arrangement between NIB and various Nordic countries had already been used for certain projects. See MD 2000/3598, m. 1, minutes from meeting held 10 November 2000 between NIB and Norwegian ministries, dated 22 November 2000.

17 Archival unit MD 2000/3598, m. 1, contains a large number of letters, memos and agreement drafts concerning the new modernization project. The following description is a summary of this documentation with emphasis on the agreements involving Norwegian agencies.

18 MD 2000/3598, m. 1, Agreement between the Kingdom of Norway and the Russian Federation concerning the modernization of Joint Stock Company Kola Mining and Metallurgy Company, signed by Norwegian minister of the environment Siri Bjerke and Russian minister of the economy German Gref, 19 June 2001.

19 Ibid., fax from Yuri Evdokimov to Norwegian General Consul in Murmansk Robert Kvile, dated 15 June 2001.

20 Ibid., email correspondence between Norwegian Ministry of Foreign Affairs and Norwegian General Consulate in Murmansk, dated 2 October 2001.

21 One may argue that the comprehensive set of agreements and guarantees surrounding the modernization at this time (see also later in this chapter) were merely expressions of NIB's usual business practices rather than a reflection of faltering trust between the two parties. However, in choosing to involve NIB to the degree that they did, the NME was well aware that a watertight legal framework would be established. This, I would argue, was an expression of the Norwegian wariness at this point.

22 A further three agreements were also concluded: between NIB and Boliden Contech, Boliden Contech and Norilsk Nikel and between NIB and the Swedish Government which sponsored the effort with USD 3 million. These, however, were not the object of Norwegian scrutiny, and are therefore not reviewed in the present study. In addition to these agreements came the loan contract regulating payments, interest rates and repayments between NIB and KGMK.

23 MD 2000/3598, m. 1, Grant Facility Agreement between Nordic Investment Bank and Joint Stock Company Kola Mining and Metallurgical Company, dated 17 December 2001.

24 Ibid., m. 1, Support, Share Retention and Guarantee Agreement – Grant Facility Agreement between Nordic Investment Bank and Joint Stock Company Mining and Metallurgical Company Norilsk Nikel, dated 17 December 2001; MD 2005/4143, status report from NIB, dated 4 May 2007.

25 Ibid., m. 2, Pechenganikel Smelter Modernisation Programme: Technical Annex, dated and signed 20 December 2001.

26  Ibid., m. 1, Agreement between the Kingdom of Norway and the Russian Federation concerning the modernization of Joint Stock Company Kola Mining and Metallurgy Company, signed by Norwegian minister of the environment Siri Bjerke and Russian minister of the economy German Gref, 19 June 2001, Article 7.2; ibid., Agreement between the NME and the Nordic Investment Bank concerning management of funding for the modernization of Kola Mining and Metallurgy Company, signed 19 December 2001, Article 8.2. Titles of both agreements are translated from Norwegian.

27  *Aftenposten*, 23 September 2002.

28  Bond and Levine, 'Noril'sk Nickel'.

29  MD 2000/3598, m. 2, letter from Romanov to the Norwegian General Consulate in Murmansk, dated 3 October 2002.

30  Ibid., letter from NME to the Norwegian General Consulate in Murmansk, dated 10 October 2002.

31  Ibid., letter from Romanov to Brende, dated 16 October 2002. See also email from Norwegian Consulate-General in Murmansk, containing report from meeting with Romanov, dated 17 October 2002.

32  Ibid., fax from Burukhin to NIB and Boliden Contech, dated 18 October 2002.

33  The Norwegian support package was geographically limited to the Pechenga area and could not be transferred to a modernization in Monchegorsk, further to the southeast. According to the legal framework, all Norwegian funding would be reimbursed if the money was not spent according to the original project plan. See MD 2000/3598, m. 2, internal memo (AIP), dated 17 June 2003.

34  Ibid., internal memo (AIP), dated 25 October 2002.

35  Ibid., internal memo (AIP), dated 17 June 2003.

36  Ibid., letter from Børge Brende to Mikhail Prokhorov, dated 6 November 2002.

37  See for example ibid., Report to donors from NIB, 2 February 2003; ibid., Report to donors from NIB, 29 July 2003; ibid., Report to donors from NIB, 28 February 2004; MD 2005/4143, Report to donors from NIB, 20 September 2005.

38  MD 2000/3598, m. 2, Report to donors from NIB, 2 February 2003.

39  A detailed and technologically informed description of the modernization project can be found in MD 2000/3598, m. 2, Pechenganikel Smelter

Modernisation Programme: Technical Annex, dated and signed 20 November 2001.

40 MD 2000/3598, m. 2, fax from NIB to NME, dated 29 July 2003.

41 Ibid., memo (AIP), dated 15 August 2003.

42 Ibid., letter from NME to NIB, dated 19 August 2003.

43 Ibid., letter from NIB to NME, dated 13 February 2004; ibid., memo (AIP), dated 20 February 2004.

44 This example is from ibid., Report to Donors from NIB, dated 28 February 2004. See previously cited examples earlier in this chapter and also MD 2005/4143, Report to Donors from NIB, dated 20 September 2005; ibid., Report to Donors from NIB, dated 7 July 2006.

45 For an entertaining account of the squabbles between various owner interests in Norilsk Nikel in the late 2010s and the instability in concern management that ensued, see Stephen Fortescue and Vesa Rautio, 'Noril'sk Nickel: A Global Company?' *Eurasian Geography and Economics* 52, no. 6 (2011): 835–56.

46 MD 2005/4143, Report to Donors from NIB, dated 20 September 2005.

47 Bond and Levine, 'Noril'sk Nickel'.

48 MD 2005/4143, Report to Donors from NIB, dated 20 September 2005.

49 MD 2005/4243, letter from NIB to NME, dated 16 January 2006.

50 MD 2005/4143, memo (AIP), dated 30 January 2006.

51 MD 2005/4143, internal memo (dept. for organization and economy, NME), dated 17 February 2006. In this document, the intention behind NME's probing into the legal aspects of the delays seems to point in this direction: 'On this background [the delays] the AIP-department suggests . . . that the Attorney General of Civil Affairs is asked . . . to evaluate the conditions for nullifying the agreement with NIB . . ., which in turn would entail that NIB would have to nullify its agreement [with KGMK] concerning financial contribution to the modernization.'

52 MD 2000/3598, m. 1, Agreement between the NME and the Nordic Investment Bank concerning management of funding for the modernization of Kola Mining and Metallurgy Company, signed 19 December 2001, Article 8.2.

53 MD 2005/4143, letter from the Government Attorney's Office to NME, dated 17 February 2006.

54 Ibid., letter from NME to NIB, dated 21 February 2006.

55  See for example MD 2005/4143, memo (AIP), dated 6 March 2006, where it is stated that 'there are rumors on political level in Finnmark that the Russians are not going to go through with the modernization and that Norway might be tricked with regard to funding'.

56  Ibid., minutes from meeting between NME and NIB, 9 February 2007, dated 12 February 2007.

57  Ibid., letter from NME to NIB, 9 March 2007.

58  Some examples from the Norwegian press: *Aftenposten*, 16 February 2007; *Bergens Tidende*, 19 February 2007; *Aftenposten*, 23 February 2007; *Nordlys*, 27 February 2007; *Adresseavisen*, 28 February 2007; *Nordlys*, 9 March 2007.

59  The argument for an adjustment in Norway's Russia policy, incorporating the example of the Pechenganikel modernization, was made by the present author in *Aftenposten*, 26 February 2007. This argument is further developed in Hønneland and Rowe, *Nordområdene – hva nå?*, pp. 35–55.

60  MD 2005/4143, email from the Norwegian embassy to Russia to various Norwegian governent agencies, dated 11 March 2007.

61  Apparently, the visit had been arranged at the behest of Norilsk Nikel and Yevdokimov. Within the Ministry of the Environment, the assumption was that NIB's probing in Norilsk Nikel's decision making on Pechenganikel's future had led the company and the governor to wish to meet with the Norwegians. See MD 2005/4143, minutes from meeting between Norilsk Nikel, NME and NIB 19 April, dated 24 April2007.

62  Ibid.

63  Ibid.

64  Ibid., minutes from meeting between Governor Yevdokimov, KGMK, Russia's embassy to Norway, the Norwegian Ministry of Foreign Affairs and NME and NIB 20 April, dated 24 April 2007.

65  For the 2000 assessment, see earlier in this chapter and MD 1997/2749, m. 10, letter from SFT to NME (AIP), dated 16 August 2000.

66  MD 2005/4143, letter from NIB to NME, 4 May 2007.

67  *Aftenposten*, 27 July 2007; *Nordlys*, 23 July 2007.

68  MD 2005/4143, letter from NIB to NME, dated 2 June 2008.

69  MD 2005/4143, memo (AIP), dated 6 June 2008.

70  Ibid., memo (AIP), dated 26 February 2009.

71  Ibid.

72  This and the following section are based on MD 2005/4143, minutes from meeting between NME, the Norwegian Ministry of Foreign Affairs, the Norwegian Ministry of Finances, the Norwegian Pollution Control Authority and the Attorney General of Civil Affairs, dated 14 October 2009. In addition, my discussion has been informed by a conversation with a high-ranking Norwegian official who has been involved in the modernization efforts since the very beginning in the late 1980s.

73  These improvements would come in addition to those Norilsk Nikel, according to stated company plans, would implement unilaterally. The NME was informed of such plans and did expect them to come to fruition. This detail was given in conversation with the author by NME official.

74  MD 2005/4143, letter from the Norwegian Ministry of Foreign Affairs to NME, dated 9 November 2009; ibid., letter from the Attorney General of Civil Affairs, dated 10 November 2009. The following discussion is partially based on conversation with a well-informed individual (see previous footnote).

75  MD 2005/4143, letter from NME to NIB, dated 22 November 2010.

76  Unless otherwise stated in a reference, the following presentation is based on information received during a conversation with a high-ranking official who was involved in the modernization efforts for many years.

77  MD 2005/4143, letter from Minister of the Environment Erik Solheim to the Storting Standing Committee on Energy and the Environment, dated 23 November 2009.

78  Another comparison has a similar result: Relative to discharges from the operations of the International Nickel Company of Canada, which is of similar size, discharges from Norilsk Nikel subsidiaries are about five times more. Supposedly, the main reason is outdated technology in the Russian factories. See Vesa Rautio, *The Potential for Community Restructuring: Mining Towns in Pechenga* (Helsinki: Kikimora Publications, 2004), p. 41.

79  Office of the Finnmark County Governor, 'Pasvik Programme Summary Report: State of the Environment in the Norwegian Finnish and Russian Border Area', Report 1/2008, Vadsø: Office of the County Governor of Finnmark, Dept. of Environmental Affairs, 2008, pp. 6–7 and p. 20.

80  Press release from the NME, 'Norilsk Nikel oppfyller ikke moderniseringsavtale', dated 11 November 2009. On file with the author.

81  *Finnmark Dagblad*, 20 December 2009.

82  MD 2005/4143, memo (AIP), minutes from meeting between the Russian Ministry of Natural Resources/ the Russian embassy to Norway and NME, 18 February 2010, dated 26 February 2010. In the minutes, the stipulation of the 2001 agreement is mistakenly quoted as a discharge of 20,000 tons annually. The upper emission limit that was referred to by the Norwegians, the so-called 'critical load', was a measure established internationally in the LRTAP process. The most common definition of the concept is that given by scholars Nilsson and Grennfelt: 'A qualitative estimate of an exposure to one or more pollutants below which significant harmful effects on specified sensitive elements of the environment do not occur according to present knowledge.' See J. Nilsson and P. Grennfelt (eds.), 'Critical Loads for Sulphur and Nitrogen', *UNECE/Nordic Council Workshop Report*, Copenhagen: Nordic Council of Ministers, 1988. The concept is not applied to Russian environmental policies, as Russia has not ratified the second sulphur protocol to the LRTAP Convention, in which the 'critical load' concept was introduced.

83  MD 2005/4143, memo (AIP), minutes from meeting between the Russian Ministry of Natural Resources/the Russian embassy to Norway and NME, 18 February 2010, dated 26 February 2010.

84  Here the Norwegians were probably referring to the PRC concept, which they had funded with 30 million NOK. This assumption is strengthened by the fact that, shortly after the meeting, the NME drafted a letter to their Russian counterparts offering to forward technical documentation from PRC (MD 2005/4143, draft letter from NME state secretary Heidi Sørensen to Russian deputy minister of natural resources, attached to memo dated 10 March 2010).

85  MD 2005/4143, memo (AIP), minutes from meeting between the Russian Ministry of Natural Resources/the Russian embassy to Norway and NME, 18 February 2010, dated 26 February 2010.

86  Arguably the two foremost ministers of the Stoltenberg government had been directly involved in the modernization saga in the early stages of their political careers. During his time as state secretary (junior minister) in the Ministry of the Environment, Stoltenberg held several talks with

Russian counterparts on the matter. Equally significantly, Jonas Gahr Støre, the Norwegian minister of foreign affairs since autumn 2005, had worked intensively on the Pechenganikel issue while he served as adviser to the Syse government in the early 1990s. Gahr Støre has regularly brought the issue up in talks with Russian colleague Sergei Lavrov.

87 From 'Norsk-russisk felleserklæring fra statsministeren i Kongeriket Norge og presidenten i Den russiske føderasjon', dated 27 April 2010. On file with the author.

# Chapter 6

1 Darst, *Smokestack Diplomacy*, p. 49.

2 Ibid., pp. 50–2.

3 Ibid., pp. 125–7.

4 Olsen, 'Moja po tvoja?', pp. 77–9. Olsen argues that these Russian 'tactics of the vise' were not without precedence. Referring to the 1977 'gray zone' negotiations between Norway and the Soviet Union, Olsen sees the Soviet approach to consultations as similar to that found in the Pechenganikel negotiations and suggests that this is a more general Russian pattern.

5 See Bronder et al., 'Norilsk Nickel'. The Norwegian approach to the Norilsk Nikel concern had at this time changed. The Norwegian government support for Bellona in this case must be understood as part of a greater effort to counter attempts from Norilsk Nikel at 'greenwashing' the concern's public image, that is, trying to appear more environmentally minded than it was. This interpretation was confirmed in conversation with a Norwegian government official.

6 The leader of Bellona Foundation, Frederic Hauge, reiterated this view in a radio debate in 2007.

7 In the bilateral environmental cooperation with Russia, one of Norway's main targets was to 'develop environmental competencies in Russian industry, public management bodies and science'. See Utenriksdepartementet, 'Om handlingsprogrammet for Øst-Europa', Oslo: Utenriksdepartementet, 1994–5, p. 19.

8 Oldfield, *Russian Nature*, p. 130.

# References

Aanesen, Margrethe (2006). 'To Russia with Love? Four Essays on Public Intervention under Asymmetric Information: The Petsjenganikel Case on the Kola Peninsula', Ph.D. dissertation, Norut report 08/2006. Tromsø: Norut Samfunnsforskning AS.

Andresen, Steinar, Elin Lerum Boasson, and Geir Hønneland (2008). 'Framveksten av internasjonal miljøpolitikk', in Steinar Andresen, Elin Lerum Boasson, and Geir Hønneland (eds.), *Internasjonal miljøpolitikk*. Bergen: Fagbokforlaget.

Andresen, Steinar, Elin Lerum Boasson, and Geir Hønneland (eds.) (2012). *International Environmental Agreements*. New York: Routledge.

Åslund, Anders (1998). 'Russia's Financial Crisis: Causes and Possible Remedies', *Post-Soviet Geography and Economics*, Vol. 39, No. 6, pp. 309–28.

Åslund, Anders (2004). 'Russia's Economic Transformation under Putin', *Eurasian Geography and Economics*, Vol. 45, No. 6, pp. 397–420.

Åtland, Kristian (2003). 'Russisk nordområdepolitikk etter den kalde krigen: Forholdet mellom næringsinteresser og militærstrategiske interesser', *FFI-report 2003/00713*. Kjeller: Norwegian Defence Research Establishment (FFI).

Blakkisrud, Helge (2006). 'What Is to Be Done with the North?', in Helge Blakkisrud and Geir Hønneland (eds.), *Tackling Space: Federal Politics and the Russian North*. Lanham, MD: University Press of America.

Blejer, Mario (1998). 'Financial Crisis in Russia: A Comment', *Post-Soviet Geography and Economics*, Vol. 39, No. 6, pp. 329–31.

Bond, Andrew R. (1996). 'The Russian Copper Industry and the Noril'sk Joint-Stock Company in the Mid-1990s', *Post-Soviet Geography and Economics*, Vol. 37, No. 5, pp. 286–329.

Bond, Andrew R. and Richard M. Levine (1989). 'Air Pollution Problem at Monchegorsk Continues', *Soviet Geography*, Vol. 30, No. 3, pp. 255–61.

Bond, Andrew R. and Richard M. Levine (2001). 'Noril'sk Nickel and Russian Platinum-Group Metals Production', *Post-Soviet Geography and Economics*, Vol. 42, No. 2, pp. 77–104.

Bronder, Larisa, Alexander Nikitin, Kristin V. Jorgensen, and Vladislav Nikiforov (2010). 'Norilsk Nickel: The Soviet Legacy of Industrial Pollution', *Bellona Report 2010*. Oslo: Bellona Foundation.

Bruno, Andy Richard (2011). 'Making Nature Modern: Economic Transformation and the Environment in the Soviet North', Ph.D. dissertation in history, University of Illinois, Urbana, IL.

Bruno, Andy (2016), *The Nature of Soviet Power: An Arctic Environmental History*, Cambridge: Cambridge University Press.

Brusletto, Hanne (1994). 'Forhandlinger mellom Norge og Sovjetunionen om kraftutbygging i Pasvikelven 1945–1963: Norsk–sovjetisk brobygging under den kalde krigen', MA thesis in history, University of Oslo, fall 1994.

Castberg, Rune (1993). 'Felles problem – ulik prioritering: nordisk–russisk miljøsamarbeid og nikkelverkene på Kola', *Nordisk Østforum*, Vol. 2, pp. 15–24.

Cohen, Stephen F. (2000). *Failed Crusade: America and the Tragedy of Post-Communist Russia*. New York: Norton.

Darst, Robert G. (2001). *Smokestack Diplomacy: Cooperation and Conflict in East-West Environmental Politics*. Cambridge, MA: MIT Press.

Ellingsen, Arvid (1997). 'Aksjon "Stopp dødsskyene fra Sovjet" – en umulig lederoppgave', Masters thesis, University of Bergen, fall 1997.

Fortescue, Stephen (2006). *Russia's Oil Barons and Metal Magnates: Oligarchs and the State in Transition*. Basingstoke: Palgrave Macmillan.

Fortescue, Stephen and Vesa Rautio (2011). 'Noril'sk Nickel: A Global Company?' *Eurasian Geography and Economics*, Vol. 52, No. 6, pp. 835–56.

Fylkesmannen i Finnmark (Office of the County Governor of Finnmark) (1987). 'Norsk/Sovjetisk møte om miljøvern i felles grenseområder, Kirkenes 17.–19. juni 1986'. Vadsø: Fylkesmannen i Finnmark.

Galushko, O. Ya. (1985). '40 let trudovoy vakhty gorno-metallurgicheskogo kombinata "Pechenganikel"', *Tsvetnaya Metallurgiya*, Vol. 9, pp. 1–3.

Gerner, Kristian (1993). 'Naturmiljø och politik i Österuropa' *Nordisk Østforum*, Vol. 2, pp. 5–14.

Gerner, Kristian and Lars Lundgren (1978). *Planhushållning och miljöproblem: Sovjetisk debatt om natur och samhälle 1960–1976*. Stockholm: LiberFörlag.

Goldman, Marshall I. (1972). *The Spoils of Progress: Environmental Pollution in the Soviet Union*. Cambridge, MA: MIT Press.

Goldman, Marshall I. (2004). 'Anders in Wonderland: Comments on Russia's Economic Transformation under Putin', *Eurasian Geography and Economics*, Vol. 45, No. 6, pp. 429–34.

Goldstein, Judith and Robert O. Keohane (1993a). 'Ideas and Foreign Policy: An Analytical Framework', in Judith Goldstein and Robert O. Keohane

(eds.), *Ideas and Foreign Policy: Beliefs, Institutions and Political Change.*
Ithaca, NY: Cornell University Press.

Goldstein, Judith and Robert O. Keohane (eds.) (1993b). *Ideas and Foreign Policy: Beliefs, Institutions and Political Change.* Ithaca: Cornell University Press.

Gorbachev, Mikhail S. (1988). *Izbrannye rechi i stati*, vol. 5. Moscow: Izdatelstvo politicheskoi literatury.

Hønneland, Geir (2003). *Russia and the West: Environmental Co-operation and Conflict.* London: Routledge.

Hønneland, Geir (2005). *Barentsbrytninger: Norsk nordområdepolitikk etter den kalde krigen.* Kristiansand: Høyskoleforlaget.

Hønneland, Geir (2010). *Borderland Russians: Identity, Narrative and International Relations.* Basingstoke: Palgrave Macmillan.

Hønneland, Geir and Anne-Kristin Jørgensen (2003). *Implementing International Environmental Agreements in Russia.* Manchester: Manchester University Press.

Hønneland, Geir and Lars Rowe (2004). *Health as International Politics: Combating Communicable Diseases in the Baltic Sea Region.* Burlington, VT: Ashgate.

Hønneland, Geir and Lars Rowe (2005). 'Western versus Post-Soviet Medicine: Fighting Tuberculosis and HIV in North-West Russia and the Baltic States', *Journal of Communist Studies and Transition Politics*, Vol. 21, No. 3, pp. 395–414.

Hønneland, Geir and Lars Rowe (2008). *Fra svarte skyer til helleristninger: Norsk–russisk miljøvernsamarbeid gjennom 20 år.* Trondheim: Tapir Akademisk Forlag.

Hønneland, Geir and Lars Rowe (2010). *Nordområdene – hva nå?* Trondheim: Tapir Akademisk Forlag.

Kärnä, Veikko (2007). 'A Return to the Past? An Institutional Analysis of Transitional Development in the Russian Mining Industry'. Turku: Publications of the Turku School of Economics.

Kolstø, Pål (1990). *Kjempen vakler: Sovjetunionen under Gorbatsjov.* Oslo: Universitetsforlaget.

Komarov, Boris (1980). *The Destruction of Nature in the Soviet Union.* White Plains, NY: M. E. Sharpe.

Korostelev, Aleksansdr (2008). *Delo 'Norilskii Nikel': Privat-kapitalizm Rossii. Strategiya, taktika i metody promyshlennoi privatizatsii po Chubais. Priroda proiskhozhdeniya oligarkhicheskikh kapitalov.* Moscow: Algoritm-Kniga.

Kotkin, Stephen (2001). *Armageddon Averted: The Soviet Collapse 1970–2000*. Oxford: Oxford University Press.

Kotov, Vladimir and Elena Nikitina (1993). 'Russia in Transition: Obstacles to Environmental Protection', *Environment*, Vol. 35, No. 10, pp. 10–20.

Kotov, Vladimir and Elena Nikitina (1998). 'Implementation and Effectiveness of the Acid Rain Regime in Russia', in David G. Victor, Kai Raustiala, and Eugene B. Skolnikoff (eds.), *The Implementation and Effectiveness of International Environmental Commitments: Theory and Practice*. Cambridge, MA: MIT Press.

Leopold, Aldo (1949). *A Sand County Almanac*. Oxford: Oxford University Press.

Matsak, V.A. (ed.) (2005). *Pechenga. Opyt kraevedcheskioi entsiklopedii*. Murmansk: Prosvetitelskiy tsentr 'Dobrokhot.'

Neumann, Iver B. (1994). 'A Region-Building Approach to Northern Europe', *Review of International Studies*, Vol. 20, No. 1, pp. 53–74.

Nilsson, J. and P. Grennfelt (eds.) (1988). 'Critical Loads for Sulphur and Nitrogen', *UNECE/Nordic Council Workshop Report*. Copenhagen: Nordic Council of Ministers.

Office of the Finnmark County Governor (2008). 'Pasvik Programme Summary Report: State of the Environment in the Norwegian Finnish and Russian Border Area', *Report 1/2008*. Vadsø: Office of the County Governor of Finnmark, Dept. of Environmental Affairs.

Oldfield, Jonathan D. (2005). *Russian Nature: Exploring the Environmental Consequences of Societal Change*. Aldershot: Ashgate.

Olsen, Gro Elisabeth (2005). 'Moja po tvoja? Vilkår for samarbeid med Russland: en studie av Petsjenganikel-prosjektet 1985–2002', MA thesis in history, University of Bergen, spring 2005.

Pharo, Helge Ø. (2008). 'Reluctance, Enthusiasm and Indulgence: The Expansion of Bilateral Norwegian Aid', in Helge Ø. Pharo and Monika Pohle Fraser (eds.), *The Aid Rush: Aid Regimes in Northern Europe during the Cold War*. Oslo: Unipub.

Pryde, Philip R. (1991). *Environmental Management in the Soviet Union*. Cambridge: Cambridge University Press.

Putnam, Robert D. (1988). 'The Logic of Two-Level Games', *International Organization*, Vol. 42, No. 3, pp. 427–60.

Rautio, Vesa (2000), 'Petsamo – "Kaipaukseni maasta" globaalitalouden pyörteisiin', *Terra*, Vol. 112, No. 3, pp. 129–40.

Rautio, Vesa (2004). *The Potential for Community Restructuring: Mining Towns in Pechenga*. Helsinki: Kikimora Publications.

Rotihaug, Ingunn (2000). "'For fred og vennskap mellom folkene": Sambandet Norge–Sovjetunionen 1945–70', *Defence Studies* 1/2000. Oslo: Norwegian Institute for Defence Studies.

Rowe, Elana Wilson (2009). 'Who Is to Blame? Agency, Causality, Responsibility and the Role of Experts in Russian Framings of Global Climate Change', *Europe-Asia Studies*, Vol. 61, No. 4, pp. 593–619.

Rowe, Lars (2002). '"Nyttige idioter": Fredsfronten i Norge, 1949–1956', *Defence Studies* 1/2002. Oslo: Norwegian Institute for Defence Studies.

Rowe, Lars (2020). *Industry, War and Stalin's Battle for Resources: The Arctic and the Environment*. London and New York: I.B. Tauris.

Rowe, Lars and Geir Hønneland (2010). 'Norge og Russland: Tilbake til normaltilstanden', *Nordisk Østforum*, Vol. 24, No. 2, pp. 133–47.

Rowe, Lars, Geir Hønneland, and Arild Moe (2007). 'Evaluering av miljøvernsamarbeidet mellom Norge og Russland', *FNI Report* 7/2007. Lysaker: Fridtjof Nansen Institute.

Sakwa, Richard (1989). *Soviet Politics: An Introduction*. London: Routledge.

Sakwa, Richard (2002). *Russian Politics and Society*, third edition. London: Routledge.

Service, Robert (2005). *A History of Modern Russia from Nicholas II to Vladimir Putin*. Cambridge, MA: Harvard University Press.

Shulman, Marshall D. (1963). *Stalin's Foreign Policy Reappraised*. Cambridge, MA: Harvard University Press.

Skedsmo, Pål (2010). *Russisk sivilsamfunn og norske hjelpere*. Trondheim: Tapir Akademisk Forlag.

Smirnov, M.B. (ed.) (1998). *Sistema ispravitelno-trudovykh lagerei v SSSR*. Moskva: Zvenya.

Solvang, Ola (1998). *Under de tre pipene: Fotografier fra Nikel*. Tromsø: Bladet Nordlys.

Stortingsproposisjon nr. 1 (1991–1992). Budsjetterminen 1992 (Norwegian National Budget for 1992). Oslo: Stortinget.

Stortingsproposisjon nr. 1 (2008–2009). Budsjetterminen 2009 (Norwegian National Budget for 2009). Oslo: Stortinget.

Stortingsproposisjon nr. 1 (2010–2011). Budsjetterminen 2011 (Norwegian National Budget for 2011). Oslo: Stortinget.

Svirskii, Grigorii (1960). 'Komandirovka v Nikel', *Ogonek*, Vol. 24, pp. 20–1.

Syse, Else, Christian Syse, and Henrik Syse (eds.) (2003). *Ta ikke den ironiske tonen: Tanker og taler av Jan P. Syse*. Oslo: Press Forlag.

Utenriksdepartementet (1994–1995). 'Om handlingsprogrammet for Øst-Europa'. Oslo: Utenriksdepartementet (Ministry of Foreign Affairs).

Volkov, Vadim (2003). 'Obshchestvennost': Russia's Lost Concept of a Civil Society', in Norbert Götz and Jörg Hackmann (eds.), *Civil Society in the Baltic Sea Region*. Aldershot: Ashgate.

Weiner, Douglas (1988). *Models of Nature: Conservation and Community Ecology in the Soviet Union, 1917–1935*. Bloomington, IN: Indiana University Press.

Weiner, Douglas R. (1999). *A Little Corner of Freedom: Russian Nature Protection from Stalin to Gorbachëv*. Berkeley, CA: University of California Press.

# Index